THE RICHES BENEATH OUR FEET

THE RICHES
BENEATH OUR FEET

How mining shaped Britain

GEOFF COYLE

OXFORD
UNIVERSITY PRESS

OXFORD
UNIVERSITY PRESS

Great Clarendon Street, Oxford OX2 6DP

Oxford University Press is a department of the University of Oxford.
It furthers the University's objective of excellence in research, scholarship,
and education by publishing worldwide in

Oxford New York

Auckland Cape Town Dar es Salaam Hong Kong Karachi
Kuala Lumpur Madrid Melbourne Mexico City Nairobi
New Delhi Shanghai Taipei Toronto

With offices in

Argentina Austria Brazil Chile Czech Republic France Greece
Guatemala Hungary Italy Japan Poland Portugal Singapore
South Korea Switzerland Thailand Turkey Ukraine Vietnam

Oxford is a registered trade mark of Oxford University Press
in the UK and in certain other countries

Published in the United States
by Oxford University Press Inc., New York

British Library Cataloguing in Publication Data

Data available

Library of Congress Cataloging in Publication Data
Library of Congress Control Number: 2010920359

Typeset by SPI Publisher Services, Pondicherry, India
Printed in Great Britain
on acid-free paper by
Clays Ltd., St Ives Plc

ISBN 978–0–19–955129–3

1 3 5 7 9 10 8 6 4 2

PREFACE

This book outlines the history of how mining people, and the industries they toiled in, have shaped Britain over many centuries, the two main features of this story being the longevity and diversity of British mining. For instance, copper mining in North Wales goes back to the Bronze Age and went on for thousands of years, as did tin and copper mining in Cornwall and Devon. What are now the beautiful Yorkshire and Derbyshire Dales were scarred by lead and silver mining from Roman times until the twentieth century. Coal was mined by the Romans but it was not until the early seventeenth century that it became significant, eventually growing to be the biggest mining sector of all, employing a million men at its peak. Iron mines in Cumbria and Cleveland fed the shipyards and steelworks of Barrow-in-Furness and the Durham coast towns, and those of the Midlands were the basis for great industries in Birmingham and Sheffield. The fields around Rievaulx Abbey in North Yorkshire are now verdant and dense with sheep, but 700 years ago they were a hive of smoking industry as the monastic community ran mines and smelters, the wealth from which supported the beauty, charity, and worship of the Cistercian communities. The oldest of all is flint mining, which started at least 4,500 years ago. There is *much* more to Britain's mining history than those illustrations, but the mines produced the metals and coal that provided our material wealth, and supported Britain's eventual expansion to a great manufacturing and trading nation. The source of all this wealth was Britain's surprisingly varied geology.

Although we tend to regard mining as taking place underground, we have to cover Britain's other extractive industries (excluding oil and gas) and describe quarrying for stone and granite, and open-cast iron mining in, for instance, Lincolnshire. 'Britain' means England, Scotland, and Wales; the main minerals mined in Britain and their uses are described in a ready-reference appendix at the end of this volume. Ireland has its own mining history which is deserving of a book in its own right.

The history of mining is also the tale of its people. We shall look at how the miners themselves lived, worked, and all too often died, and see how women once slaved underground in coal mines, even with their young children. In the lead mines of the Pennines women often ran small farms while the men and older boys wrought in the mines. The mine owners are fascinating subjects, and not always the hard and rapacious exploiters one might expect.

Mining benefited enormously from the very inventive people who created the machines that made large-scale mining possible, and the work of the miner less arduous and dangerous, and we'll explore that history. Mere words can't do justice to the industries, their people, and the legacy in the landscape, so 16 pages of photographs bring those things to light. There are a few maps of the mining districts, some diagrams, and an extensive bibliography.

I have been almost overwhelmed by the sheer volume of existing literature dealing in detail with particular industries, or even individual mines, and one of the enjoyable challenges of writing this has been to take sources of 300 pages or more on given topics and to bring out their main themes in a few hundred words. Those studies are the efforts of people who have done all the painstaking work on local records to show what happened in, for instance, Cornwall's copper and tin mines or South Yorkshire's coal mines. I am deeply grateful to them, because, without their work, this story of British mining for the general reader could not have been written in a lifetime. Unfortunately, in a book of a reasonable size we cannot cover every facet of all these industries. Experts in particular aspects of mining history will find that much detail has had to be omitted, and I extend my apologies for that.

One of the delights of writing this has been the electronic acquaintanceships that I have made with helpful and friendly people in museums and mining companies. Without their cooperation and constructive comments the book would have died at birth. I cannot name them all, and to mention some but not others would be ungracious. The downside to that has been a few organizations who have simply not replied to polite letters and emails asking for information.

The background to this book is that many years ago I chose to go to the Royal School of Mines to study mining engineering, mainly because of the variety of subjects that were covered: geology; surveying; mechanical, civil, and electrical engineering; mining methods; and finance and mine valuation were among them, all supported by a healthy dose of mathematics. As part of the course I worked underground in coal mines in Lancashire and Yorkshire and at a copper mine in Sweden, and made numerous visits to mines in, for instance, Kent and Cornwall.

After graduation, as part of a training programme for aspiring colliery managers, I worked 2,300 feet down a coal mine in Lancashire, and I even have a certificate to say that I am qualified to work on a coal face. Unhappily, even in the 1960s, coal mines in Lancashire were closing and there were few prospects for a young engineer when so many much more experienced people were looking for jobs. However, my university tutor—a marvellous man called Alan Grierson who subsequently became the Crown Estates Mineral Agent—had always said that a mining engineer can turn his hand to anything, so I heeded his advice and did other things with my career. Despite that, I have never lost my fascination for this amazing industry and the people who made it.

No author writes a book without a good deal of help from others, in this case my very tolerant editor at Oxford University Press, Latha Menon, and all the other people there who piloted the book through publication, the endlessly helpful and patient staff at the Faringdon branch of Oxfordshire County Libraries, and the people who kindly reviewed my drafts. My dear wife, Julie, encouraged me to do this, supported me while I did it (as she has done in all my work), and criticized my drafts. Any errors of fact or interpretation are entirely my own.

If, on reading this, you are inspired to explore further, the bibliography may be helpful. In addition, bodies such as the Open University organize geological field trips to various parts of the country. There are also a number of volunteer groups engaged in research into local mines. This involves either tracing and, in some cases preserving, underground or surface workings, or detailed archaeological study of surface features such as pumping and winding engines. You will need to be reasonably fit, properly equipped, and, above all, willing to take instructions from a more experienced member of the group. Mines, especially old ones, can be dangerous places. Should you come across what seems to be an old mine on no account should you attempt to explore it without expert supervision. There are grave risks of falling down an abandoned shaft or being trapped and seriously injured—even killed—by a rock fall.

Preface

Finally, I hope that you will enjoy reading this book as much as I have enjoyed writing it. If it has excited you about this remarkable part of Britain's history it has achieved its purpose.

RGC

Shrivenham

CONTENTS

Contents

LIST OF FIGURES

LIST OF PLATES

1

The Achievements of Britain's Miners

The Old Testament's Book of Job elegantly describes the task of the miner (Chapter 28, verses 9 to 11, King James version):

> He putteth forth his hand upon the flinty rock; he overturneth the mountains by the roots. He cutteth out rivers among the rocks; and his eye seeth every precious thing. He bindeth the streams that they weep not. And the thing that is hidden he bringeth forth to light.

Bringing precious things to light is easier said than done. For instance, the stone used to build King Solomon's Temple came from Solomon's Quarry under Jerusalem, which is 650 feet deep, and the miners must have worked in harsh conditions. Light came from primitive oil lamps; bronze tools were all that was available for cutting the stone; horse- or man-power dragged the blocks of cut stone and raised them to the surface. Wooden props would be used to support overhangs; the water that always gets into mines was probably carried to the surface in buckets; and ventilation must have been a very chancy thing. The injuries and deaths can only have been appalling. One can only stand in respectful amazement at such an achievement as, even with electric power, modern stone-cutting equipment, lights, ventilation fans, water pumps, metal roof supports, accurate surveying techniques, and transport and hoisting machinery, it would be quite a challenge to extract large blocks of stone good enough for the Temple, let alone to dispose of all the waste cuttings. Britain's miners and mining engineers have, though, brought to light vast quantities of coal, metals, salt, slate, and stone, and in doing so have shaped Britain's technology, its landscape, and its political and social institutions, and provided the basis of much of our wealth and ease of living. That, then, is the trail that this book traces out, through 4,500 years of mining in Britain.

For thousands of years the main sources of power in a mine for underground transport and hoisting to the surface were animals and the human body, all too often females and even children. As technology advanced, these were eventually replaced by waterwheels, steam, and eventually electricity but, astonishingly, the basic labour of the miner did not change much until fairly modern times. Until the early twentieth century the task of a collier was to lie on his side cutting a slot under the coal with a hand pick. As the slot grew deeper, it had to become higher to accommodate his shoulders and was supported on small wooden pit props, imported from the Baltic and cut to length with hand saws. When he and his mates had finished the undercutting, explosives were used to break the coal, and the next shift shovelled the coal into small railway wagons, called tubs, in which it was taken to the surface. He walked to and from the pit in his work clothes, which his wife or mother washed by hand, and he bathed in a tin tub in front of the fire. Fortunately, by the 1960s things had improved somewhat. The coal was undercut by a machine, rather like an electric chainsaw with a blade 4 feet long, that was dragged along the coal face by a built-in cable but, after the blasting shift, the stint was to shovel 16 tons of coal onto a conveyor belt. There were also pit-head baths with separate rooms for clean and dirty clothes and showers in between, although these improvements came very late in mining's long history.

Coal can be hacked out with a hand pick but the metals—tin, copper, lead, zinc, etc.—are in hard rock. Nevertheless, even in rock as hard as sandstone or granite, the job of driving tunnels and excavating the ore was, for millennia, done by hand using picks, hammers, and other tools that we'll meet in due course, though rates of progress in a tunnel were, at best, only a few yards in a month. Explosives were not used until about the sixteenth century but that required shot-holes drilled by hand. In a confined, ill-lit space one man held a long 'boring bar' while another used a heavy hammer to drive it to the required depth in rock as hard as granite. That was bad enough in a relatively horizontal tunnel, but when the holes had to be overhead the task beggars belief. The early black-powder explosive left appalling fumes in what were then badly ventilated mines but, since the men were paid by the distance tunnelled or the ton mined, time couldn't be wasted waiting for all the fumes to clear. It was not until the 1880s that pneumatic rock drills and dynamite explosive became available, and even then it took years for these to become universal. Bear in mind, though, that pneumatic rock drills are bigger and heavier than the drills used to dig up a road and need a pneumatic 'leg' to support them.

Nowadays, really advanced technology has, as we'll see, made the miner's task less laborious and dangerous, and greatly increased his daily output. We will come to the numerous developments of mining technology in due course. First, let us go back in time and look at the origins of mining in Britain in Neolithic times, and touch on the development of mining in copper, tin, lead, iron, coal, and all the other substances that have been brought into the light, before looking at each one more closely in subsequent chapters.

Roughly 10,000 years ago in the Near East people who had lived a precarious existence as hunter-gatherers started to settle in fixed places and to farm: the 'new stone age', or Neolithic period, had begun, though this process itself took some considerable time, only reaching Britain in about 4,000 BC. This was a period of great, if probably slow, social and technological change. Three factors interacted, though we cannot say that one was more important than another—they seem to have occurred more or less simultaneously, over a long period of time. For convenience, we can say that the first was the ability to accumulate and preserve surplus food to support specialist workers, such as priests and miners. Secondly we must recognize considerable sophistication in society that allowed powerful people to allocate the surplus to the specialists and assemble large numbers of people for major projects. Finally, growing more food calls for tools to clear land by felling trees, and weapons for continued hunting of meat and to defend surplus food and the winter stock. This is where flint mining comes into the picture, and the products of the flint mines enabled more land to be cleared and defended, yielding ever greater prosperity. This technological explosion took place in Britain about 4,500 years ago; mining is a very ancient profession.

Flint is a very hard form of silica—silicon dioxide—sometimes black, but with a wide range of colours, especially when wet. It is formed from the silica in the skeletons of many millions of creatures, such as sponges, which fell to the floor of an ancient ocean. Flint occurs quite widely in Britain, wherever the land is what geologists call Upper Chalk, such as in Kent, Sussex, and Wiltshire, mainly in southern England. The flint is normally in bands or layers in the chalk, but where the chalk has weathered, lumps of flint can be found on the surface. Palaeolithic people discovered that the lumps, when hit with another stone, would split along clear fracture lines and, carefully worked, could take an edge as sharp as a modern razor. (Obsidian, volcanic glass, is even sharper but is rare. It was highly valued by Neolithic people and was extensively traded.) Modern experimental archaeologists have revived the ancient skill of flint knapping and have proved that a flint blade, attached to a shaft of wood, makes an axe that will fell a small tree just as a modern steel axe will, but more slowly. Flint does not decay or erode, so if, with the landowner's permission, you walk over a field in a flint area, looking closely at the ground, you have a fair chance of finding a flint tool, though it may have been broken by ploughing.

Flint had, of course, been used for many millennia for hunting and, probably, fighting implements, but the Neolithic demand for flint called for greater output and a more refined and varied range of products. Searching for greater quantities of flint, and of better quality, led to the discovery of the flint layers. Where the layer was near the surface, scraping away the overlying chalk must have given a good yield of flint. The layers also often show on the edge of a chalk cliff, from which the flint could be obtained by quarrying, though a lot of chalk has to be shifted to get a fairly small quantity of flint, and increased output required going deeper underground. The archaeological site of Grimes Graves in Suffolk, dated to between 2500 and 2300 BC, and which is nothing to do with burials, shows numerous pits, dug down to about 30 feet, which we can fairly say were the first mines in Britain.

Figure 1 gives an idea of what was involved. The shaft was dug using an antler as a pick, which would also be used to dig the flints. The shoulder bone of an ox would serve as a shovel and the flints and waste chalk would have to be hoisted to the surface by a rope made of animal hide or on the backs of people. Access to the pits was by

The Achievements of Britain's Miners

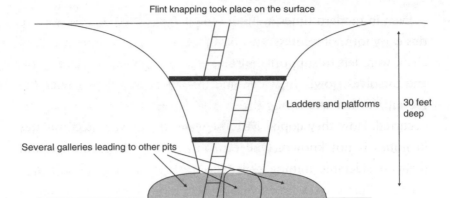

FIG 1 Flint mining.

ladder, probably via an intermediate platform of logs; a usable ladder can be made by cutting steps into the side of a tree trunk. No food remains have been found in the pits, so the miners probably came to the surface for 'lunch'. There may have been a village nearby for the miners and their families, or mining may have been a seasonal activity, probably in summer when fine weather made working the pits more feasible and it was easier to transport the product by packhorse. Lighting came from animal fat burning in a hollow cut into a lump of chalk.

It was a big industry; the site, which is open to the public, covers 34 acres and about 350 pits have been identified, so the pits were about 45 yards apart. At the bottom of the pit, tunnels radiated out to follow the flint layer, or seam, eventually connecting with other pits to maximize recovery of the valuable flints. As each pit was worked out, the rubble from the next pit was dumped into it, both to leave easy access to the site and to maintain ground support.

The Achievements of Britain's Miners

Even in modern times, mining can be dangerous, but the risks in this early form of mining must have been shockingly high. Pillars of chalk were left to support the roof, even though chalk is fairly soft and dissolves slowly in water, and tree roots would penetrate the ground, further weakening the overlying rock. Collapses must have occurred. How they coped with the water that always accumulates in mines is not known. One can only admire their courage and their considerable mining skill. The product of their labours was used in chisels, axes, saws (made by gluing chips of flint into a piece of wood), and all manner of tools for skinning animals, preparing the hides for use as clothing, cutting and dressing meat, shaping wood, and, of course, spears and arrows for hunting and fighting. With these tools, land could be cleared and defended, so we may reasonably say that flint mining helped to shape Britain. Similar flint mines have been found at Cissbury, Blackpatch, and Harrow Hill in Sussex, and elsewhere.

From the late Middle Ages flint, carefully shaped into blocks (a very considerable skill), was used to build houses in Sussex, Kent, and elsewhere, providing some very attractive buildings. The blocks cannot be made as exact as brick, so brick or stone had to be used for door and window frames and for wall corners. Many of these houses are now listed buildings, so flint blocks have to be used for repairs, and the art of block-making is not dead. In more recent times, flints were used to fire guns, even into the nineteenth century—the Battle of Waterloo in 1815 was fought with flintlocks. Flints were used for a thousand years or more as a source of fire, as flint strikes sparks from a piece of steel. Crushed flint has been used for road-making. Nowadays, flint is used for gravel drives, and roasted and finely ground flint is used in the pottery industry. Since flint is very hard it can also be used in grinding mills for soft rocks. While flint is useful it has limitations—flint implements are laborious to make and the cutting edge breaks easily—and was supplanted for utensils, tools, and weapons by metals, such as copper, tin, and iron. The value of these metals in enabling civilization to evolve is incalculable, and eventually produced great mining industries that flourished all over Britain until relatively recent times, often leaving evidence in the landscape, notably the stark beauty of the remains of mines and processing mills that make the coast of north Cornwall so atmospheric.

Flint mining dates from antiquity and is long extinct, apart from small quantities for building restoration, but tin and copper were mined in Britain continuously for about 3,500 years, and lead and gold for at least two millennia. The sheer longevity of these mining activities is one of their many remarkable features. In later chapters we'll explore in detail the histories of the copper, tin, and iron industries, and others, such as coal, lead, zinc, salt, slate, china clay, and stone, tracing their origin, growth, and eventual decline. We'll look at changes in mining technology but also at the lives of the people, social change, and the evidence in the landscape that these industries once flourished. For now let me just indicate the vast scale of the mining industries of Britain by a few examples: Cornwall and parts of Devon had thousands of tin and copper mines at one time or another, from ancient times until the last one closed in 1998. In the 1860s, the output of tin metal was around 10,000 tons a year; in the late eighteenth century two copper mines in North Wales were the world's main producers of that metal; the Cleveland district of North Yorkshire had some 80 ironstone mines; the biggest of all was, of course, the coal industry, which in 1870 produced half the world's coal. It remained a huge industry so that, when the collieries were nationalized in 1947, about a thousand coal mines were taken into public ownership. As late as the 1990s there were 70 pits in South

Yorkshire employing 115,000 men, women having long been banned from working underground in coal mines (that practice was almost unknown in metal mines); and slate from North Wales was (and still is) of such high quality that it was exported, for example, to provide the roof of the Argentine National Bank in Buenos Aires, as well as other notable buildings worldwide.

And that is without even mentioning salt, china clay, stone for buildings and roads, or limestone for making steel and concrete. Britain was fortunate in being blessed with so many mineral riches and with the skills, courage, and ingenuity of its people, and the creativity to develop the technology to exploit them.

We can show this in another way in a map (Figure 2) of the main mining areas of the British mainland, though to identify each and every one would be impossibly complex. But wherever you live in Britain you will not be very far from some indication of mining, past or present, or from a mining museum.

The downside to this record of success was, unhappily, the loss of miners' lives, the crippling accidents and dreadful illnesses, allied to the effects on the miners' wives and often large families. We will meet these sad stories in more detail as we explore the separate industries. Suffice it to say that, for example, in 1911 alone, 2,000 British coal miners were killed and 160,000 injured. The collieries at the time employed about a million men, so this is a high casualty rate.

From time to time I'll need to mention historical costs, prices, and incomes. It is very difficult to express these in modern terms in any meaningful way. For instance, in the 1860s the Duke of Portland's annual income from royalties on coal mined from the rich seams under his estate amounted to £100,000 or more. We could estimate that as the equivalent of £20 million in current money. On the other hand, a miner's wages were about £50 a year, which,

FIG 2 Some of Britain's mining fields (excluding coal and main lead/zinc areas).

on the same basis of comparison as the Duke, would now be £10,000 per year. Nowadays, the Duke would still be an unbelievably rich man, but the miner's wages would be less than half of average earnings, or little more than the state retirement pension for a married couple. The modern conversions do no more than re-emphasize the incredible disparity, so in general I shall quote costs, prices, and incomes in the money of the time and leave you to draw your own conclusions. Converting into decimal currency is necessary, but when we say that a miner's wife had to keep a family of nine on 16 shillings (80p) a week, we scarcely show how hard that was. We'll also need to mention distances and thicknesses; in general I'll use feet and inches for measurements, but I'll use metres when we look at scientific geology in Chapter 2. For place names I have used only those to be found in a contemporary road atlas, such as *Collins*. Occasionally I have had to simplify; some mining villages were too small to be shown on maps, in which case I have referred to the nearest identifiable place.

Before we look further at how the vast mining industries developed, we'll discover where the minerals come from by covering some aspects of British geology, one of which is that what is now southern England was once near the Antarctic Circle. Thereafter we can trace the stories of the main metals, coal, salt, and stone. Before that, let me mention what seem to me to be some remarkable coincidences.

We are sometimes led to believe that prehistoric people were ignorant, shambling primitives, clad in animal skins and eking out a precarious existence. This probably is far from the truth precisely because their existence could be precarious at times, they needed to be intelligent and very observant of their environment, with highly retentive memories and good (sometimes remarkable) organizational ability. The building of Stonehenge and similar structures could not have happened without the ability to accumulate and preserve big surpluses of food and to organize a large workforce over a very considerable period of time. It also needed high levels of skill to manoeuvre and erect stones weighing many tons; even today no one is quite sure how it was done, though there are many theories. To the skills of organization and technology we have to add a good knowledge of geography. The very large 'sarsen' stones at Stonehenge had to be moved over hilly ground for some miles, and the bluestones come from the Welsh mountains. That must have involved movement along the coast of South Wales and up rivers, suggesting knowledge of boat-building beyond simple canoes, and of navigation techniques, though an alternative (but less plausible) theory is that they were transported by ancient glaciers, so-called erratics or rocks that do not originate where they are found. Avebury is in many ways even more evocative than Stonehenge. In Scotland, the Isle of Barra in the Outer Hebrides is 50 miles off the mainland, and the seas are, to say the least, not always tranquil. Nonetheless, Barra was inhabited from about 4,000 years ago, and there is evidence that ox-drawn ploughs were used. That raises the questions of how people got to Barra in the first place, and how large and intractable oxen were transported there. Even if they came south from the larger Hebridean isles, that makes one wonder how they got across the straits of the Minches. These were remarkable achievements.

So, while these ancient people have left no written records, we can be sure that they were very far from being stupid. Let us look at some of the things they had to work with. The most abundant substances available to them were the power of the sun; the oxygen in the air; animals such as deer, and the food, skins, bones, and sinews they provide; clay; timber; and various forms of rock, including those we now recognize as containing common metals such as copper, tin, and iron. Iron is, after aluminium, the most common metal, making up about 4.6 per cent of the Earth's crust.

We cannot know how metals were first discovered, but a reasonable conjecture is that stones used to contain a fire were seen to melt in the heat, the stones in question being the result of weathering of the overlying rock, usually limestone, or broken rocks washed down by rivers from outcrops of the ore.

Here are the remarkable coincidences:

Bricks, made from clay dried in the sun, can be used to line pits for storing food reserves; timber, burning in oxygen, can produce a high enough temperature to smelt a metal such as copper; by limiting the flow of air, suitable timbers can be converted into charcoal, which gives an even hotter flame, capable of smelting iron; blowing on a fire makes it burn brighter, so it is only a small step to making bellows from animal skins to produce sustained higher temperatures in a simple furnace; clay can be used to make moulds in which metals can be cast in useful shapes; clay, heated in a furnace, becomes hard enough to be a plate or a cooking pot, and the plate will be tough enough for prolonged use, even standing up to wear from utensils made from the metals.

The melting point of copper is 1083°C, and that of iron is 1538°C, but had Nature contrived these to be even one- or two-hundred degrees higher it is hard to imagine how civilization could ever have got beyond flint technology.

The fact that ancient peoples noticed, experimented with, and exploited these coincidences (if they are coincidental?) was an astounding accomplishment. We may certainly feel that they laid the first foundations for our modern lives.

2

Britain's Geology and Minerals

Britain has a remarkably varied geology for such a small island, and this has given rise to the country's very wide range of mining industries. For instance, there are large deposits of salt under Cheshire, formed by the evaporation of sea water over millions of years when Britain—or what eventually became Britain—was a very hot place. There are vast amounts of limestone, formed from the shells of marine creatures, so we were once under a warm, shallow sea; and we still have large reserves of coal that came from tropical forests and swamps. Fossils show that dinosaurs once roamed what is now Britain, and the plant fossils that are seen with the dinosaur remains are large ferns, so the climate in those times was much warmer than it is now. There is also evidence in British rocks of Sahara-like deserts, sub-tropical reefs, tropical rain forests, and cold or temperate conditions. So, there is a fascinating story beneath our feet of how these varied rocks came to exist and what has been happening to Britain during that time.

The task of the geologist is to unravel that story, not only for its intrinsic interest but also to aid in the search for metals, coal, and other valuables. We can see the geology below the ground in mine workings, but the deepest British mines only extend about 5,000 feet below the surface, and most are shallower. While those first few thousand feet are important to the mining engineer and the miner, exploration geologists need to know about the deeper rocks and how and when they were formed, as that information may help to locate additional mineral resources. They see their role as being the study of the physics and chemistry of the planet. Fortunately, patient tracing of the sequence of different types of rock over many miles gives a very good idea of how these rocks are related in depth and over long periods of time, and what happened to them.

Close examination of particular places also reveals a great deal; Lulworth Cove in Dorset is obviously a bay being scooped out by the sea, but behind it the rocks are folded and crumpled. The rocks are limestone and contain fossils of marine creatures, so these rocks were once on the floor of a warm sea but have been uplifted to the surface. Vast forces must have acted to distort them to such an extent. At the other end of the country, Fingal's Cave on the island of Staffa in the Hebrides has beautiful hexagonal columns of very hard volcanic basalt sandwiched between earlier and later volcanic flows. Those flows did not form columns because the lava contained dissolved gases, as shown by careful examination of the site and by studying very thin slices of the rocks under the microscope. Geologists supplement their fieldwork by laboratory analysis of the chemistry of rocks, and rock samples can be subjected to very high temperature and pressure to simulate the conditions that are thought to create mineral deposits. Such work has shown why Welsh slate splits so cleanly as to provide very high-quality roofing material.

The ages of these rocks were originally estimated by looking at how thick a layer of sandstone was, how long it took to build up an inch of silt in a slow river, and dividing one by the other. To the early geologists, such as Adam Sedgwick (1785–1873), the results were staggering—even shocking—as they suggested that Earth was unbelievably older than the creation date of 4004 BC calculated in the seventeenth century by Bishop James Usher from the Bible genealogies. The calculation from geology was a very rough estimate, and could only be used on some types of rock, so early geologists had to rely on *relative dating*, the sense that if rock A is on top of rock B which is on top of C, then C is the oldest and A is the most recent. That can get complicated when there are many rocks in a sequence, not just three, and the rocks are folded and crumpled, as at Lulworth and many other places. Very detailed relative ages can be worked out by studying the fossils in a sequence of rocks using the techniques of palaeontology and micropaleontology. In fact, a lot of careful study is often required to enable geologists to unravel the geological history of a particular area.

However, about 100 years ago a new science—geochronology—developed and was able to give absolute dates for rocks so that we can now say that rock A is X million years old. The technique is *radiometric dating*, which works on the same principle as radiocarbon dating in archaeology but, instead of measuring the decay of carbon from one form into another, uses other elements such as uranium, which decays into a form of lead. Combining the relative dates, which come from study of the rocks in the field, with the radiometric dates from the laboratory has allowed geologists, after a lot of debate and discussion over the past 50 years, to work out the sequence and ages of the main systems of rocks.

It turns out that the oldest rocks in Britain are in Wester Ross in Scotland, and are 2 *billion* years old, but some of the oldest rocks on Earth are in Greenland and are twice that age. More 'recently', only about 90 million years ago, a major inflow of the sea occurred during which the remains of billions of minute marine organisms eventually deposited about 500 metres of chalk over Britain and north-west Europe. Much of this has since been eroded, though the iconic White Cliffs of Dover and the beautiful Needles off the Isle of Wight still remain. During the formation of the chalk, sponges grew on the sea bed, and the silica—silicon dioxide—in their remains is the source of the flint nodules mentioned in Chapter 1. If the use of flint started the eventual development of human society, we might say that our way of life is all down to the humble sponge.

The types and sequences of rocks are only part of the geological story, and we also need to understand why Britain, for instance, is where it is on the surface of planet. The reason why the landmasses of Earth are constantly shifting is now accepted to be *plate tectonics*. Briefly, the top 60 to 70 miles of the Earth—the *lithosphere*—is made up of about 11 large plates of rock, both on the surface and under the sea, and throughout billions of years these continental and oceanic plates have drifted very slowly due to currents in the underlying mantle produced by the Earth's internal heat. Mountains arise where the plates push together and oceans form where they pull apart. This is the process of *continental drift*, in the sense that the continents sit on these plates and move relative to one another. There are many indications of land masses that seem once to have fitted together, but the clinching evidence for continental drift comes from detailed study of changes in the Earth's magnetism which are recorded on the sea floor. Such work shows that what became England and Wales was once nearer to the South Pole than to the North Pole, with southern England positioned roughly on the Antarctic Circle, though Scotland was further north. Over hundreds of millions of years Britain drifted 12,000 km (about 7,500 miles) north, but spent about 300 million years near the equator.

As continental drift takes place molten rock, called magma, rises from the interior to form new plate material beneath the oceans.

At the same time oceanic crust gets driven back down, which causes intense volcanic and earthquake activity. To take one case, a large patch of the crust lying beneath the eastern Pacific Ocean is being slowly pushed under South America, which is the reason for the Andes mountains, the region's volcanic eruptions and earthquakes, and, indirectly, its vast deposits of copper. The Himalayas and Tibet are the result of India colliding with central Asia. Continental drift is a very slow process but can be monitored by very careful surveying, now usually done with satellite measurements. This shows that where the Eurasian plate, where Britain is—which extends from the mid-Atlantic to the Philippines—is moving away from the North American plate at about 2 cm (four-fifths of an inch) a year, which, over a human lifetime, amounts to about the height of an adult person.

In the past, Britain has had many volcanoes—the rock on which Edinburgh Castle sits is an eroded volcano. Fortunately, these volcanoes are all now extinct, but Britain gets hundreds of fairly small earthquakes each year, many of them not even noticeable except to the seismological instruments. However, in February 2008, Britain had its biggest earthquake for 25 years, measuring 5.2 on the Richter scale, centred under Market Rasen in Lincolnshire, but felt over much of the country and as far away as the Netherlands. There was widespread damage to chimneys and roofs, but fortunately only one serious injury. Except to the people concerned this was trivial by comparison with the earthquake that destroyed San Francisco in 1906. That was 8.25 on the Richter scale (which is a logarithmic scale such that each increase of 1 represents a ten-fold increase in the power of the earthquake); in other words, the San Francisco earthquake was about a thousand times as powerful as the one in Market Rasen. Britons are very lucky to live in a quiet zone.

Earthquakes are caused by slip along geological faults, where one area of rock moves up, down, or sideways relative to its neighbour in order to release stresses that have been built up by the tectonic forces. Such faults can be seen on a relatively small scale, and happening quickly, in an earthquake, but the large geological faults form over a long time span and involve movements of thousands of feet. The Great Glen in Scotland sits on an ancient fault line that runs from Fort William to Inverness, and probably continues under the sea. The Craven Fault in North Yorkshire involved a vertical shift of several thousand feet, and there are many other big faults in Britain. The bane of the miner's life is the unsuspected fault that disrupts his carefully planned programme of work by displacing ore or coal from its expected position.

How Britain Came to Exist

The Earth is about 4,600 million years old. To cope with what happened during all that time, geologists divide it into a number of periods, or systems, each of which has its own characteristic set of fossils. We won't mention all of them, but the rocks dating from the formation of the planet until 542 million years ago are loosely grouped together and referred to as the *Precambrian*. The rocks in Wester Ross are Precambrian; and if you go west on minor roads from Church Stretton in Shropshire to the village of Stiperstones, you cross the Long Mynd, which is an area of Precambrian rock in England. The Precambrian was mostly barren of complex life, though some enigmatic fossils and traces at its very end indicate that there were early forms of life by this time.

In the *Cambrian* system, which started 542 million years ago and lasted for 54 million years, a prominent life-form in the ocean was the well-known trilobite, which evolved considerably during this epoch.

The trilobite is curious and amusing but the Cambrian-era Burgess Shales in Canada contain fossils of life-forms that are out of a bad dream, beautifully described in Stephen Gould's *Wonderful Life*. During the 70 million years after the Cambrian what became southern Britain moved about 3,000 km, roughly 2,000 miles, from near the Antarctic Circle to about 30° south latitude. Scotland—at that time separate from the rest of Britain—was somewhere around the equator, though what became England and Wales eventually caught up.

The *Silurian* period dates from around 444 million years ago, and lasted 28 million years. The main event was a continental collision between two tectonic plates, which created mountains over much of Britain, which was then about 20° below the equator: coral reefs of this age have been found in the English Midlands.

The next period we'll look at is the famous *Carboniferous* system, lasting from 360 to 299 million years ago. A warm, shallow sea deposited vast quantities of calcium carbonate from the remains of plankton and eventually formed the Carboniferous limestone in many parts of Britain. When the seas were invaded by rivers from a neighbouring landmass the coarse sandstone called Millstone Grit was formed. During the later Carboniferous these rivers supported luxuriant swamps and rain forests which were repeatedly covered by more river sediments and limestone due to episodes of change in sea level. The buried peat from the forests decayed and was compressed to form coal seams sandwiched between the sandstones and shales of the Coal Measures which eventually supported Britain's huge coal industry. Carboniferous limestone, when split with a geologist's hammer to give a fresh surface, often has a slightly oily smell, due to traces of organic matter deposited with the limestones.

The *Permian* era is dated to between 299 and 250 million years ago. Britain was, by then, about 30° north of the equator, but we were in the middle of a vast super-continent which geologists call Pangaea. The climate was hot and dry, and dune fields were widespread. The end of the Permian saw a mass extinction of life, in which about 95 per cent of species died out. There has been much debate about the reason for this, and many theories have been advanced. The extinction of the dinosaurs, which happened much later—about 60 million years ago—is usually attributed to an asteroid collision, but no sign of such an impact has been seen in the Permian, though craters of such an age are not easy to find. Another possibility is that the retreat of the shoreline during the formation of the super-continent changed the climate beyond the range which the existing creatures could tolerate. A third theory is massive volcanism.

During the *Triassic* system, representing the period from 250 to 200 million years ago, Britain was at approximately 20–25° north, or about where the Sahara desert is now. Huge sand-dune fields covered vast areas of the country. Invasion by extremely salty sea water from a transient sea produced salt deposits up to 1,000 metres thick that now extend into the north of England. Those 50 million years saw the start of new reptile and ammonite species—ammonites being the wonderful spiral fossils one sometimes sees in souvenir shops, especially along the south coast—but the Triassic ended with another mass extinction of life. It is almost as though the planet has an occasional need to refresh its life-forms.

The *Jurassic* system commenced about 200 million years ago and lasted around 55 million years. The sea spread over much of Britain and the early Jurassic seas were full of marine life, such as plesiosaurs. The 'Loch Ness Monsters' have even been said to be a surviving population of plesiosaurs, although this idea is about as far-fetched as the film *Jurassic Park*. Various inundations of the sea produced limestones, and the later stages of the Jurassic saw the deposition of clays rich in organic material. One of these, the

Kimmeridge Clay, extends under the North Sea and is an important source of oil.

The *Cretaceous* system started about 146 million years ago and lasted for about 80 million years. It also ended with a mass extinction when about 75 per cent of species died off, notably the dinosaurs. One theory is that this was due to the impact of a comet or asteroid in Mexico's Yucatan Peninsula, others are that it was due to a combination of such a collision, colossal volcanic eruptions, and the long-term decline of animal species.

Recent geological time, the *Tertiary*, runs from 65 million years ago to the present. Britain had moved to its present latitude, the climate was warm and humid, and we were still attached to Europe. There was extensive volcanic activity for about 10 million years of these 65 million, the relics of which are distributed across Britain, most notably in the volcanic rocks of the Hebrides and in places such as Glen Coe in Scotland. It's quite hard to grasp these long spans of time, but 10 million years is equivalent to about 300,000 human generations. People with an interest in genealogy would be delighted if they could trace their ancestry for 10 generations.

It was only in the last few million years that our remote ancestors, the various hominids, evolved. In the last few hundred thousand years recognizable humans evolved. As the ice sheets from the most recent glaciation melted, Britain became separated from Europe, and by 5,000 years ago was about its present shape, except that East Anglia was larger and there were islands in the North Sea.

Minerals, Ores, and Types of Rock

A *mineral* is a substance with a definite chemical composition. Rocks are made up of several, or many, minerals. Nearly all minerals are chemical compounds of two or more elements, and *ore minerals* contain one or more elements of economic interest. For instance, copper, which has many uses because it conducts heat and electricity very well, may occur as copper oxide, 'cuprite', which is copper chemically combined with oxygen. Copper also occurs in many other chemical forms, such as copper pyrites, 'chalcopyrite' (copper, iron, and sulphur combined), and even sometimes as native copper, which is more or less pure copper metal. A common ore mineral of tin is cassiterite, tin oxide. Many minerals are very beautiful: chalcopyrite forms yellow/golden crystals as does pyrite, iron sulphide. Pyrite, when ground to a powder, can easily be mistaken for gold dust, and the unwary have been known to buy a 'gold' mine only to be disappointed; its common name is 'fool's gold'. Azurite, another copper mineral, is a vivid azure blue. Azurite's relative, malachite, is a valuable copper mineral sometimes displaying bright green bands which are so attractive that it is also polished as jewellery, ornaments, and table tops. The Appendix to this book lists the main ore minerals that have been mined in Britain, as a ready-reference.

An *ore* is an accumulation of ore minerals in a body of rock from which it is the task of the miner to extract them for a profit, if that can be done. In fact, 'ore' is really an economic term, as a given accumulation of ore minerals, or *mineral deposit*, may be profitable or not, depending on operating costs and the market prices of the metals that can be extracted. A case in point is South Crofty, which was the last working tin mine in Cornwall, closing in 1998. However, the market price of tin tripled between 2004 and 2007, so the mine may re-open within the next few years, this time probably using the latest mining technology. We'll meet South Crofty a few more times in this book.

Some minerals once had no known use and were a nuisance to the miner, but became valuable ore minerals when applications were eventually discovered. A good example is wolframite—a mineral of iron, manganese, and tungsten—which occurs in some Cornish tin ores, but it was not until armoured warships were developed in the mid nineteenth century that the manganese and tungsten found a use for making armour plate. Nickel, now with many valuable uses, was first discovered in the copper mines of the Harz Mountains in Germany in the sixteenth century. It gets its name from the German word 'cupfernickel', or Devil's copper, as it could not be refined in the way that copper was, and had to be discarded. In Britain, nickel occurs in ironstone in South Wales.

Unfortunately, ores inevitably contain a proportion of waste rock; I have a sample of azurite picked up on a beach in Cyprus which has the beautiful blue of the mineral interspersed with bands of limestone; Cyprus was once a major producer of copper and gets its name from the Greek word for this metal. Such a mixture of valuable mineral and waste rock is called the mineral deposit or the ore body. Mineral deposits are of all shapes and sizes. A seam of coal is usually more or less flat, though sometimes sloping at 30° or more, and usually much the same thickness over large areas. The veins, or lodes, of metal-bearing minerals, by contrast can be any size or shape. Sometimes they continue for long distances, while in other cases they peter out and have to be found again. They can range from a few inches to many feet thick and are often at steep angles. An ore body can also be very large indeed. A molybdenum mine—molybdenum is a metal used to make super-hard steels—in Colorado 12,000 feet up in the Rocky Mountains has an ore body 5,000 feet wide and at least a mile deep. It was mined by tunnelling under the ore body, drilling long holes into it, firing explosives, and allowing the broken ore to collapse so that, once the ore has been extracted through tunnels beneath the shattered ore, the remaining mountain could collapse safely onto the ore body, a process requiring exquisite care.

The valuable minerals have to be extracted from the ore coming from the mine, which contains some waste rock. A common technique is crushing or grinding it to very fine sizes, and washing it in water, relying on the fact that the metal minerals are quite dense and will settle to the bottom when washed—for instance cassiterite, tin oxide, is about seven times as dense as water. The grinding and washing is repeated as often as necessary in order to extract as much mineral as possible. After that the mineral has to be *refined* to extract the metal in a pure form, perhaps by roasting at high temperature to drive off any sulphur as sulphur dioxide, a very unpleasant gas. Refining is an environmentally unfriendly process so, in modern times, much effort is devoted to making it as clean as possible. That was not always the case, and old mining districts were often terrible places in which to live. The fumes from the refinery darkened the sky, killed nearby vegetation, caused lung diseases, or were even slow-acting toxins if they contained arsenic.

The waste rock from the ore has many names, usually in a local dialect. Some of them are *gangue, halvens, stent*, and *mullock*, as well as *waste, muck*, and other words not used in polite conversation. If the ore contains even 10 per cent of mineral, the 90 per cent of gangue has to be discarded, sometimes in vast heaps disfiguring the landscape. One such tip, a product of coal mining, caused the Aberfan disaster, the tragic story of which will be told in Chapter 11. The gangue usually contains some mineral that was not worth extracting at the time of mining or which was regarded as an impurity but, if metal prices rise in the future, it could be worth reprocessing heaps of gangue to extract those minerals, though it may be technically difficult to do so.

There are four main categories of rock which can contain ores, and they pose various challenges to the miner. *Igneous*, fire-formed, rocks are those that form from cooling and crystallization of magmas—molten rock. Granite is a common igneous rock and is easily recognized as it contains transparent crystals of quartz, flakes of shiny mica, and opaque white or pinkish crystals of a mineral called feldspar. The varying proportions of these three, and the sorts of feldspar that the granite contains, combine to give many different types and colours of granite. Granite usually forms irregular masses, which can range from a few miles across to thousands of square miles at the surface, and can be several miles deep, as in Cornwall. Because it is extremely hard, granite was once much used for building— Aberdeen is called 'the granite city'. Nowadays it is mainly used for kerbstones, for decorative facings on buildings, and for kitchen and bar worktops.

Another main category is the *sedimentary rocks*. As the name implies, most sedimentary rocks have been deposited in water as a result of the erosion of surface rocks by wind and rain, splitting in frosts, or glacier action over geological time. Because geological time is so long, sediments sometimes build up to thousands of feet thick, and these gradually compact and harden to form the rocks we now see at the surface.

Sandstone is a sedimentary rock, mainly quartz, but it can be hard and coarse, as in the millstone grit of the Pennines, or quite fine-grained and softer, such as sandstones in southern Britain that were once desert dunes. The finest-grained sedimentary rocks, called shale, or mudstone, are the most abundant sedimentary rock. Shales generally form when muds and silts are deposited by slow-moving currents and become compacted, such as on the deep ocean floor or in calm, shallow seas. Some shales contain large amounts of plant remains and may be an important source of oil.

Limestones are also a type of sedimentary rock. They are formed either by the accumulation on the sea bed, over geological time, of the hard, shelly remains of sea creatures, or by precipitation from sea water. Limestone sequences can be thousands of feet thick and erosion of the beds can have spectacular effects—limestone is slightly soluble in water, which results in the development of caves, stalactites, and limestone pavements with deep cracks, called clints and grykes.

Coal is also, in effect, a sedimentary rock, and an enormously valuable one, which was formed, as shown by the plant fossils it contains, from the remains of trees and ferns which grew in what must have been semi-tropical conditions. Coal seams are not usually more than several feet thick, and have been compacted by being buried under later deposits of sandstone. There are several varieties of coal, one being anthracite, the highest grade of coal, containing up to 93 per cent carbon. It is brilliant, shiny, black and can be carved and polished into ornaments. Anthracite burns with a pale-blue smokeless flame, but is expensive.

Very interesting things happen when igneous rocks intrude into sedimentary rocks. Often, the sedimentary rocks erode away faster, leaving the igneous rocks behind to form hill tops. A spectacular example of this is the Whin Sill in Northumberland. Bamburgh Castle and parts of Hadrian's Wall sit dramatically on top of the sill, which also extends to form the Farne Islands. The high temperatures and pressures of the igneous intrusion can also alter the surrounding local rocks into the third important class, the *metamorphic rocks*, which can sometimes be very rich in minerals.

The final rock category that is significant to mining is the *evaporites*, which are sources of salt minerals. These are formed when shallow lakes evaporate faster than the inflow of water from rainfall or rivers, which can only happen in very hot conditions. The Great Salt Lake in Utah is an example of that process. Evaporites can also be produced by evaporating sea water; in modern times this occurs in the Gulf of Suez, Corfu, and similarly hot places, and is an important source of salt.

Where Do Minerals Come From?

As we have seen, coal is the result of peat from the remains of Carboniferous forests being compressed by overlying layers of rock, and was formed between 373 and 290 million years ago. Limestone and chalk come from the shells of countless millions of small sea creatures, deposited in great thicknesses on the floor of a warm ocean. The origins of coal and limestone, which are very valuable to our way of life, are easy to understand, but we still have to explain where copper, tin, lead, and all the other metallic minerals come from. A great deal of careful research has shed light on this question. For instance, undersea cameras have shown hot, mineral-rich, fluids coming out of chimney-like black smokers on the sea bed and precipitating mounds of copper, iron, and lead sulphides, and metals have been measured in gases coming from volcanoes. This kind of research, added to very careful study of what can be seen below ground, has resulted in good understanding of the origins of metallic minerals. These are not only satisfying to the geologist but also of great practical importance to the mining engineer, as they provide good guidance in the search for minerals and in the planning of mining operations.

The ore minerals of lead and zinc, such as the common lead sulphide, or *galena*, and zinc sulphide, or *sphalerite*, are often found and mined together in one mineral deposit, often within limestone such as in the Pennines. The presence of these minerals in the compressed remains of seashells is now understood to be the result of hot fluids containing the lead and zinc minerals rising from the magma and penetrating into the limestone's natural cracks, or dissolving the limestone to create cavities, cooling and precipitating ore minerals in these spaces. Samples of azurite and malachite (copper minerals) picked up off a beach in Cyprus clearly show veins of the minerals cutting through limestone, so this so-called hydrothermal, or hot-fluid, explanation is convincing. Tin and copper are a little more complicated as they are nearly always found in granite rocks, such as in Cornwall. The theory goes that the metal minerals and water are dissolved in the igneous magma and, as it cools, the water separates from the magma, carrying the metallic minerals with it. As this fluid cools and reacts with the surrounding rock the ore minerals precipitate in fractures. Many other metallic minerals are regarded as coming about in much the same way—by replacement of the existing, or 'country', rock by hot, mineral-bearing fluids.

It's worth restating that metalliferous ore bodies are of all shapes and sizes. The most important type in British mining are veins, also called lodes, in which the ore is in sheets which are thick enough to be worth mining and which run for a reasonable distance, though often at a steep angle. Very often there are numerous veins at different depths, and the veins intersect in all sorts of complicated ways. To make matters worse for the miner, veins often run out and have to be found again. South Crofty mine in Cornwall over a period of some 400 years mined copper and tin from more than 40 lodes to a depth of 3,000 feet in an area 2.5 miles across. A variant on the vein deposits is the stockwork in which the veins are very small, perhaps only a few inches across, and extending for only a short distance, but with hundreds of these mini-veins throughout the ore body. Obviously it's impractical to mine these tiny veins one by one, and the whole ore body has to be extracted, perhaps by quarry-like open-pit methods, or by caving. The principle is that the lower costs of shifting large tonnages in bulk will help to make processing all that ore a paying proposition.

Finally, we have to mention the very important placer, or alluvial, deposits. These arise when millennia of erosion and weathering have broken into an ore body and particles of ore have been washed down a stream. Since the ore is much denser than water, the particles accumulate on the stream bed. If the river gravel is washed in a shallow metal pan ('panning') to get rid of the useless sand, a streak of ore is left. Teams of men laboured to shovel gravel from the river, crush it, and feed it into large wooden rocking cradles, fed with copious supplies of water from the river, to scale up the yield from panning. The work was brutally hard and done in just about all weathers, so these were tough characters.

3

Britain's Copper and Tin Mining

Copper and tin are essential components of modern life, but mankind's use of metals started as long ago as about 6500 BC with the discovery of copper in Anatolia in modern Turkey, though the earliest large-scale copper mines so far discovered are in Bulgaria, and date to about 5000 BC. For a very long time the use of copper and stone overlapped, copper being used mainly for decorative objects and high-status burials. However, as metallurgical techniques improved, copper could be cast into tools and weapons, and stone artefacts gradually went out of use in the Near East. Then, in about 3500 BC, it was discovered that when copper is alloyed with a small proportion of tin the result, known as bronze, is much harder than copper but, like copper, it can be worked into intricate shapes to create effective tools, kitchen utensils, decorative items, weapons, and armour. The Bronze Age had begun; and bronze items spread to Britain by about 1900 BC, probably via Spain. British tin reached as far away as the eastern Mediterranean, though it is unlikely that the Phoenicians travelled here, as was once thought, as the trade was via intermediaries. The trade routes covered great distances; jadeite, a variety of jade, which was highly prized as a status symbol, was mined in the Alps but has been found as far away as north Scotland.

Tin is a very scarce metal, amounting only to some 0.001 per cent of the Earth's crust, but its ores are relatively abundant in north-west Spain, in Brittany, and in Cornwall and on Dartmoor in Britain, which are also rich in copper and other minerals. Copper minerals are abundant on Anglesey, rather less so near Llandudno, and are also found in Cheshire, on the Isle of Man, in the Lake District, and elsewhere. The story of all this mining is truly remarkable, and includes the struggle to find workable deposits of minerals as well as the hard and dangerous work of mining them. By the eighteenth century conflict occurred between the mining people and the refiners who converted ore into metal. Mines opened, closed, and reopened as metal prices fluctuated, and fortunes were made and lost in mining companies. Inevitably, there were rogues and charlatans. Surveying techniques improved to make sure that mines did not run into the territory of others and to avoid old workings that were full of water, or the sea floor where mines went under the ocean. In Cornwall and west Devon, the miners had their own 'parliament' to regulate their activities and protect their interests. The life of the miner and his family was as hard as it was in all mining areas, but the religious revival of the eighteenth and nineteenth centuries had a major impact in Cornwall, as we'll see in Chapter 13. Great advances were made in technology, not least in the use of steam power, and are so extraordinary that they deserve a chapter of their own.

The legacy of mining is often seen in landscapes, but nowhere more dramatically than in Cornwall where, for mile upon mile, the country is starkly defined by ruined stone buildings which were once chimneys, engine houses, mills where minerals were separated from the gangue, and smelters. Parts of the area form a UNESCO

World Heritage Site, and some of the old mines are now tourist attractions with guided tours giving insights into the daily work underground and on the surface (you would be *very* unwise to go into an old mine without a guide and proper equipment). Further north, Anglesey had huge copper mines—impressive open pits remain to mark their existence—and there were lesser ones at Great Orme near Llandudno. From the early eighteenth century, Cornwall had a prosperous china clay industry, which still functions, as we'll see in Chapter 7, and the landscape around St Austell in Cornwall had large white cones of its discarded waste—though, since the Aberfan disaster, which we'll encounter in Chapter 11, some of them have been stabilized and grassed over. For now, I want to set the tone for the book by looking at the origins of British metalliferous mining, starting in north Wales.

The Great Orme mines near Llandudno certainly date from the Bronze Age; bone tools and stone hammers have been found. Many of the tools were the leg bones of cattle, used as chisels and driven into the ore with a lump of stone. Traces of firesetting have been found, which involved heating the ore face to a high temperature and then allowing it to cool, the thermal expansion and contraction making the rock much easier to work. The Bronze Age people worked open pits at the surface but also went underground to depths of 150 feet. Whether or not the Romans mined Great Orme is unclear, though they certainly mined gold and slate in North Wales.

What happened in the Middle Ages is a blank, but Great Orme revived in Elizabethan times when German miners, who were at the forefront of mining technology throughout Europe, were granted rights to work, though Parliamentary forces destroyed the works during the civil wars of the mid seventeenth century. Activity was spasmodic during the eighteenth century but started to revive from about 1800. The Great Orme mine was worked to considerable depth with the aid of steam-powered pumps, and two new mines were developed, now abandoned and beneath Llandudno town. By the early 1850s mining was declining, due to overseas competition and water problems.

The other copper area in North Wales was on Anglesey at Parys Mountain, a mile or so inland from the small port of Amlwch on the island's northern tip. The mountain is named for Robert Parys, who made his money in 1406 by collecting fines on behalf of King Henry IV from local people who had joined Glyn Dwr (Owen Glendower) in rebellion against English rule. Copper had been mined there since the Bronze Age but after the Romans very little was done until Elizabeth I brought in German miners in 1564. For many years the main activity was precipitating copper onto iron from the polluted water flowing from the hill. It was not until 1761 that copper was discovered in abundance and mining really got going to meet the demand, much of which was for sheathing the hulls of Royal Navy ships against the teredo worm that bored holes in the wood, and clinging weeds that reduced speed and manoeuvrability.

Half the mountain was owned by Sir Nicholas Bayly, and most of the other half was in shares between Bayly and the Lewis family, but the land was just poor pasture so no one had bothered much about boundaries. When the real richness of Parys was revealed, it led to legal action between Bayly, the Lewises, and other people who owned parts of the mountain, which resulted in the rise of Thomas Williams, the 'Copper King', whom we will briefly meet again in this chapter and more fully in Chapter 12 when we look at the biographies of some of the mine owners.

After the big discovery had been made in 1761, Bayly leased the mine to Roe and Company of Macclesfield, who mined copper at

Alderley Edge, now one of Cheshire's plushest towns, and operated smelters in Macclesfield. Roe really wanted to mine lead on Bayly's land in Carnarvonshire, but a condition of doing so was that the Parys mine, and the Mona mine on the eastern side of the mountain, had to be worked for 21 years. Roe achieved no great success, and work was close to being abandoned in 1781 when two miners, Jonathan Roose and Roland Price, discovered rich copper ore a few feet down, an event that was for many years celebrated with a festival. However, Roe and Co. were experienced mining engineers and unlikely to overlook something so easily found; perhaps they were not trying too hard at Parys in order to get out of the Anglesey lease. Roe left in 1785, by which time the Mona mine was in very poor condition, as the high-grade ore had been picked out. Williams none the less arranged for £61,000 to be spent between 1785 and 1788 on smelting works, roads, and the harbour. It was a colossal sum at the time, but profits from 1787 to 1793 were £71,000 as the low-grade ore could be worked cheaply by open pits. Later, some underground working began from the bottom of the pits.

Working conditions were appalling and very dangerous. The ore was broken in the pit and wound to the surface by a windlass, or jack roll, though some men worked from platforms attached to the pit side. In the heyday of Parys about 8 tons of black powder was used in a year, and the frequent blasts caused accidents, on top of which there were major collapses in the pit due to excessive blasting and rain and frost erosion. At the surface, the ore was broken down to pieces about an inch in size by women who were paid about 10 pence (4p) for 12 hours' work, their hands being protected by gloves fitted with iron rings. Children as young as eight earned a few pence a week as helpers. The broken ore was taken to Amlwch on farm carts, roasted to remove sulphur, and shipped to market from the port, though a smelter was eventually built. The port naturally generated a small ship-repair and shipbuilding industry.

Wages at Parys and some smaller mines on Anglesey were higher than for the only other employment on the island, agriculture, but lower than those of Cornish copper miners, and that, combined with the cost advantages of large-scale open-pit mining, gave Anglesey a dominant position in the copper trade, with consequences that we shall see below. Mining on Anglesey continued on a small scale until the 1920s, but may be due for revival (see Chapter 14).

Let's now turn to the discovery and early mining of tin in Cornwall and west Devon. For many centuries tin was the south-west's main mineral product. Unfortunately, no one knows how cassiterite, tin oxide, was first discovered, though small pieces of it, called shoads, could be seen on the surface. Small quantities of gold occur with cassiterite and, as both gold and tin oxide are very dense and tend not to be washed away by surface water, it is conceivable that the tin ore was discovered in the search for gold. What is even more intriguing is how the Cornish people realized in the second millennium BC that cassiterite was very valuable. We can only speculate, but the knowledge is likely to have come via Spain, which had large tin mines.

To get at more tin than the limited amounts from shoading, trial pits, or costeans, were sunk to quite considerable depths, which implies powerful organization and great persistence. Stream tin was discovered in layers sometimes many metres thick but averaging 0.45 metres over the mining fields as a whole, laid down in valley bottoms during the last glacial period by ice erosion of tin lodes, and covered by deep layers of silt when the ice sheets melted around 12,000 years ago. The tin in the lodes from which the stream tin came had been created about 300 million years ago by hydrothermal processes within the granite of Cornwall, Bodmin Moor, and Dartmoor.

Over millennia there were many changes in the method of working the stream tin, but one, typical of medieval times and involving considerable organization, was to dig a trench, anything up to 50 feet deep, down the valley and then to work along it, shovelling the cassiterite-rich silt into a copious stream of water, brought in ditches over the moor or even from a river diverted for the purpose. Gangs of as many as 50 men worked together, and several teams might work over a number of miles down the valley, even as far as the river mouth. This was hard work, but a perquisite for the streamers was the finding of an occasional flake of gold, or even a small nugget. Each man carried a goose quill fitted with a wooden plug to hold his finds, and sold them to a goldsmith when he had enough.

At convenient points, the ore was treated on a buddle, which was initially a box or trough set at a slight angle but, by the sixteenth century, had become a large circular frame made of wood and driven by a waterwheel or horse. Its purpose was to separate the cassiterite from the gangue using copious flows of water, and relying on the fact that cassiterite has a much greater density than the silty gangue, which was washed away. All the rubble and silt was eventually washed into the sea, where it created serious problems in the harbours and fishing ports, so much so that in 1356 the Duke of Cornwall ordered streaming to stop for a while, and there was further disruption in 1532. Even after the true tin streams were exhausted, tin streaming continued to the end of Cornish mining, as cassiterite could still be recovered from the waste left behind by processing done at the mines themselves.

From pre-Roman times the cassiterite was exported to Gaul via an island called 'Ictis', which may have been St Michael's Mount or

Mount Batten near Plymouth. Such an arrangement would have made it easy to accumulate stocks until the next traders arrived.

By about the mid sixteenth century the best stream tin was exhausted and underground mining started, costeaning being used to trace back to a lode, though the early mines penetrated only to relatively shallow depths because of water inflows which were not fully over-come until the invention of steam engines in the eighteenth century. The first lodes to be extracted could be seen from the sea in the cliffs, so at Geevor, paths were constructed down into the 'zawns', natural clefts in the rocks from which tunnels, or 'adits', were driven into the lode. This was difficult labour and highly dangerous. In other loca-tions adits were driven in from hillsides, or shafts were sunk, and from about 1700, deep mines proliferated throughout south-west Britain. As there were no proper records before the mid nineteenth century, it is impossible to say how many came and went over the following 250 years, but in the St Just area alone there seem to have been more than 2,000 different shafts, adits, and surface workings.

To develop the underground tin mines groups of local men of property combined to provide the necessary funds. The expression used at the time was to 'adventure' one's capital, so inevitably they were referred to as adventurers. We shall meet some of them in Chapter 12. Without the adventurers, serious mining in Cornwall and on Dartmoor could not have got going, but the drawback was that they used the 'cost-book' system. This involved regular 'calls' from investors in proportion to their initial investment, and if a call was missed, the shares were forfeit. At the end of the year, all profits were distributed, so the cost-book system scarcely made for the long-term planning and investment that large-scale mining requires, though the problem was to some extent ameliorated by the involvement of larger-scale capitalists such as the major landowners, or mineral lords as they were called, and London-based syndicates. The eventual outcome was a huge mining industry covering much of Cornwall and west Devon, producing tin or copper, sometimes both at different times, as well as lead, iron, manganese, tungsten, and antimony, though tin and copper were dominant. The main mining districts are shown in outline in Figure 3, all of these areas having numerous mines.

Where the adventurers did not own the land to be mined, as was usual, they negotiated access from the 'mineral lord' or his agent through a system of mining setts. There were often restrictions on the setts, such as a limit of time or a requirement that work actually be done. A portion of the tin, the 'toll tin', was due to the agent, and

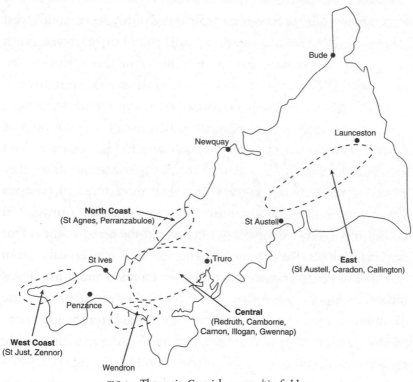

FIG 3   The main Cornish copper/tin fields.

another, 'farm tin', to the owner of the land, and there was a tax, called tin coinage, payable to the Exchequer, on refined tin before it could legally be sold. Not surprisingly, the tin coinage was particularly resented by the independent-minded Cornish mining people, and it was abolished in 1837.

Many mines were given names such as Wheal Ellen. 'Wheal' means work in a local dialect, and the name of the owner's wife or daughter has been attached as a gesture of affection or possibly as a good-luck charm. Other names, among many others, were Wheal Prosper and Wheal Prudence. A mine might have several successive names as ownership changed and amalgamations took place. Out of this multitude of mines we can mention only a few, but Cornish mining historians have described many more in detail.

Dolcoath mine, near Redruth, was one of Cornwall's richest. It was first recorded in 1738 but may date from the late sixteenth century. It was a deep mine quite early in its life—by 1780 it was down to some 600 feet, and in 1758 had one of the earliest steam engines in Cornwall. Great quantities of copper ore were produced, but the mine was closed between 1787 and 1799 due to fierce competition from Anglesey—the so-called 'copper war'. Copper reserves were exhausted by the 1840s, by which time the mine was 1,500 feet deep. However, the adventurers were sure that tin ore lay even deeper, which proved to be the case; and in 1876 it was said of it that 'the deeper it goes, the richer it gets', the main lode being about 18 feet thick. During 1871 Dolcoath's share price went from £125 to £300, but that was followed by decades of decline. The mine was constrained by old workings spread over a wide area, with no less than 70 miles of underground tunnels, and had largely exhausted its known reserves, finally closing in 1921 in the economic slump that followed the First World War.

South Crofty mine at Camborne was first worked in the 1590s under the name of Penhellick Vean. It started as a shallow tin mine, then worked deeper copper, and finally exploited the even deeper tin lodes. In all, about 40 lodes were worked, over a considerable area and down to about 3,000 feet. Rock temperatures increase with depth, the so-called geothermal gradient, and in the granite of Cornwall the temperature increase is quite rapid: jets of water from the roof at South Crofty would have made very satisfactory hot showers. In the first part of the nineteenth century, South Crofty produced tin, arsenic, and tungsten, but from the 1960s tin was the only product. In 1985 the price of tin dropped sharply, sounding the death knell of the few remaining mines, and South Crofty was the last working mine in Cornwall when it closed in 1998. The mine still has large reserves, and if the price of tin stabilizes around a sufficiently high level, it is possible that it will reopen.

Some Cornish mines, such as Geevor and Botallack in the St Just district, are perched on the very edges of the cliffs and, like many other coastal mines, worked under the sea. Both were large mines with numerous engine houses and shafts, reflecting the amalgamations of groups of adjacent smaller mines.

Carclaze mine is a few miles from St Austell and was an open pit because the ore body was a stockwork—a mass of tiny veins. The mine worked for about 400 years, producing steady, but fairly small, amounts of tin and copper, until larger amounts of tin were discovered, after which it became the largest open-pit tin mine in the world, at a mile across and 150 feet deep. The price of tin collapsed in the 1860s, which could have finished Carclaze, but happily there were large amounts of china clay. The mine's cone of white waste is one of the exceptions to the smoothing and grassing of other cones.

The products of the mines were used in very many ways. Tin and copper are versatile metals that can be alloyed to make bronze (copper with some tin), brass (copper and zinc), and pewter, which is mainly tin but with some copper and lead, though modern pewter is tin and a small percentage of antimony. Bronze resists corrosion and can be cast without voids, which made it good for cannon, while its low friction is an advantage for shaft bearings. Brass was used for household utensils, and old brass kettles and candlesticks are now valued as ornaments. Plates and candlesticks were made from pewter. Bronze, brass, and pewter are, though, rather broad-brush terms, as, depending on the precise mix of the alloy, their properties can be varied considerably, giving them even more applications: church bells, for example, are cast from a type of brass called bell metal. The tin-coated iron food can was invented as early as 1810, and by 1813 canned foods were being produced for the British Army. Copper had many uses and, as electrification became widespread, it became vital to modern life—for example, as the basis of modern domestic plumbing. New uses of copper are quite radical. The alcohol-based hand wash used in hospitals kills the MRSA virus, though not the more dangerous *Clostridium difficile*, but gel containing copper kills both. Perhaps even more significantly, harmful organisms can survive for three days on stainless-steel surfaces, such as operating tables, but for only 90 minutes on copper plate. We shall look at the potential for a revival of British copper and tin in the final chapter.

All these uses have made the mines of south-west England and North Wales immensely valuable, but the industry had what we would now call structural weaknesses in that there were many mines, but far fewer smelters. In a very real sense the smelters had the whip hand and, as cassiterite is black but refined tin is white, there was conflict in Cornwall between black and white tinners. Tin was smelted in Cornwall and smelters could become rich: indeed, Barclays bank was founded on smelting profits. South Wales became dominant for copper smelting, as coal was abundant.

The end result was that by the 1780s the situation in the industry was very unsatisfactory. Thus Thomas Williams, whom we have seen coming to fame on Anglesey, negotiated production quotas for copper, including an agreement with the smelters, which gave him effective control of the copper industry, both mining and manufacturing. The aim was to bring about some stability and to increase prices. Cornwall got the raw end of this deal with the county's annual output limited to 3,000 tons, causing mine closures and leading to great distress and serious rioting. The monopoly lasted only for four years, but it did, perhaps, bring an element of common sense and mutual benefit to what had been cut-throat competition.

The copper and tin mines were subject to boom and bust ('bonanza' and 'barrasca', as the miners called them), which happened at particular mines or across whole swathes of activity, and for all sorts of reasons. Perhaps a mine might run out of ore, or a previously rich vein might peter out and take too long to find again, throwing men out of work. A rock fall might damage the pumps, flooding the mine. There have even been hints that a mine with poor results stopped pumping in order, out of spite or envy, to flood a more prosperous neighbour—many mines were interconnected, either by design or via old workings.

To give one instance of boom and bust, Wheal Vor on the south coast of Cornwall mined very rich ore, and by 1840 employed more than 1,200 people, including about 600 women and children. Unhappily, the mine was closed between 1848 and 1853 due to a legal dispute, leaving those people unemployed. The mine became flooded during the stoppage, and when it reopened even a huge steam engine could not drain it completely. Pumping costs were £10,000 a month, but the attempt to revive the mine had to be abandoned after £300,000 had been spent. Some miners were re-employed to retrieve the shaft fittings against rapidly rising water. All told, the loss after dismantling the mine was more than £200,000, which was the same as profits had been between 1820 and 1845.

These were local events, affecting only one or a few mines, but powerful forces worked in the wider economy. The great depression of the 1840s brought widespread distress leading to the first big wave of emigration in which miners sought work elsewhere, sometimes within Britain but to a much greater extent in the United States, Australia, South Africa, Canada, and practically anywhere where jobs were on offer; it is a truism that most of the world's mining areas have people of Cornish descent. The copper slump of 1866–7 created another surge of emigrants, and a steady drain continued for many years which made it difficult to recruit men during the periodic booms. In many cases families had to be left behind in the hope of eventually rejoining the men, but those remaining behind sometimes had to survive on very little.

The big blows came from competition from places such as Malaya (as it then was), Bolivia, Australia, and other countries in which major deposits of copper and tin had been discovered. The USA had enormous copper deposits which were eventually the death knell of copper mining in Cornwall. Malaya, a British colony in the nineteenth century, was found to have very large deposits of alluvial tin which were being worked on a large scale by gangs of Chinese labourers. This used essentially the same process as Cornish tin streaming, but was much cheaper despite the transport costs, and this was a severe blow to Cornwall. However, by the early 1870s conflict between the Chinese secret societies that controlled the mine labour had reached the verge of civil war, which British troops had to suppress. That faraway trouble created a short-lived tin boom in Cornwall.

A further big disaster was the discovery in the 1870s of very rich alluvial and lode tin in Australia, which was then shipped to Britain as ballast in the wool clippers, i.e. virtually free transport. Access to this new supply of cassiterite strengthened the position of the smelters and gave them even more of a whip hand over the Cornish mines, with the result that in 1876 and 1877 132 mines out of 230 closed.

Ironically, a further blow to copper and tin came with the Companies Act of 1862. The limited liability this introduced meant that bankruptcy could lose a shareholder the value of his shares, but not everything he owned, as could happen with a cost-book partnership. While limited liability was a boon to business and commerce generally it was harmful to mining, as speculators soon abounded. The London syndicates who had financed earlier mining were knowledgeable people, but shares could now be bought by others who knew nothing about mining but who sought an income or pension, and such people were readily defrauded. Stock-jobbing, the selling of worthless stocks in sometimes non-existent mines, flourished for a while. Ridiculous promotions asserted that large pieces of tin could be picked up from the surface. In other cases there was conflict of interest in which someone promoted a mine which was known to be of no value, simply in order to make money by supplying the steam engines. R. M. Ballantyne, in his novel *Deep Down*, called such a fraud 'Wheal Do'em'. At South Wheal Leisure near Perranporth men were hired and the engine boiler was lit only when 'upcountry' shareholders were to visit. As case law under the Act evolved, these abuses were eventually stamped out, but not before some lost their money.

Mining also had political consequences, arising from the fact that miners tend to be independent-minded people, and as the distance from the seat of power in London made them difficult to administer, in many areas such as the Forest of Dean and the lead-mining districts, local customs and 'laws' evolved. This was particularly so in the tin-mining areas in Devon and Cornwall, called stannaries from *stannum*, Latin for tin, but in Cornwall the need for 'laws' was also fed by the strong sense of Cornish independence. The outcome was that, in 1197, King Richard I granted the right to stannary parliaments. This enabled the tin workers to have their civil and criminal cases heard in their own stannary courts, theoretically over riding any other legislation; in fact the stannary laws are the country's oldest written statutes, predating Magna Carta. Initially there were four stannaries in Cornwall which covered the whole of the county, later supplemented by another four in Devon, which were limited to the tin areas.

Things went further when, in 1337, King Edward III created the Duchy of Cornwall for his eldest son, Edward, the Black Prince. Since then, successive Princes of Wales have held that title and been, as it were, the monarch's deputy for Cornwall, with some responsibility for the stannary courts, though the executive role was assigned to a Lord Warden of the Stannaries, one of whom was Sir Walter Raleigh. The present duke is Prince Charles and the income from the duchy estate is used to meet the costs of his public and charitable work, and to support his family.

As the tin mines declined in importance during the eighteenth and nineteenth centuries, so the power of the stannary parliaments decayed, and the last Cornish one convened by the Lord Warden met in 1752, Devon's having ceased in 1748. The office of Lord Warden still exists as a ceremonial role and, in theory, the Duke of

Cornwall or the monarch could summon a stannary parliament, though it seems unlikely to happen. A small group of Cornish people pushing for greater autonomy for the south-west have attempted to revive the stannary concept in opposition to legislation from Westminster, but they have not as yet been successful.

We have seen what the men and women of North Wales and the south-west of England achieved, but to fully grasp the magnitude of that achievement we need to look at the work of the miner. Put simply, there are two stages to mining: getting to the ore body and then mining it: access and extraction. For access, adits (the name comes from Latin, *aditus*, an approach) are very common in steep-sided valleys such as in parts of Cornwall, the Pennines, the Forest of Dean, and many other places, and they also drain water from the mineral deposit. Vertical or inclined shafts were also sunk to reach the working levels and provide hoisting facilities, as well as to con-tribute to ventilation. A mine surveyor has to ensure that the adits and shafts go to the correct points in the mine, and also has to avoid the mining operations running into old workings that might be full of water, or into noxious gases. In 1893, 20 miners were drowned at Wheal Owles near St Just in Cornwall when a surveying error led to water breaking in, filling the shaft from 720 feet to 180 feet in 20 minutes. The pumping plant was completely overwhelmed, resulting in closure of the mine. The bodies could not be recovered.

Cornish surveying and tunnelling reached their peak in Cornwall's Great County Adit which was started in 1748 by John Williams, the manager of the Poldice mine near Redruth. The adit was begun on a tributary of the river Fal and was driven inland to reach the mine, which it did in the late 1760s, some 20 years later. It was extended to drain other mines and was finally completed in 1792, when it was 38 miles long and handled 13 million gallons (about 6,000 cubic metres) per day from a total of 40 mines. It is one of the longest water tunnels in the world—some claim it to be *the* longest—and even now it would be a remarkable feat of engineering and surveying to connect all those mines in the right places and at the correct depths. It is also a tribute to the foresight, persuasiveness (one can almost hear the cries of 'you can't do that; it's never been done before!'), and financial ability of Williams and all the others involved. Unfortunately, by 1876 the adit had been neglected, and heavy flooding released thousands of tons of silt and rocks from the adit into the river, so much so that navigation to Devoran harbour was never fully restored.

In a mine, adits are slightly inclined to provide drainage but often meander to follow the lode, in the hope of extracting enough ore to defray the cost. Many early adits were roughly pear-shaped: wider at the top than the bottom, which had small ledges on each side. After shot-holes had been driven by two men with a boring bar and heavy hammer and the shots fired, one man stood on the ledges, before the fumes had fully cleared, and raked the broken ore between his feet into baskets which a relay of boys carried to the surface for processing. The waste left over shows up as spoil tips which, even though grass has grown over them, are easily seen.

Driving adits or sinking shafts was pitifully slow. With hand drills and black-powder explosive progress in an adit might be as little as 9 feet in a month—an average of a few inches for a day of arduous labour—and cost as much as £17 per foot. When pneumatic drills and dynamite progressively came into use from about 1869 the rate increased to 2 feet per day and costs came down to £3 per foot, and that was only one of the technological advances of the nineteenth century. We'll see many more in the next chapter.

Most of the mineral lodes in metal mines are at very steep angles of 70° or more, and the ore is mined in areas called stopes which are effectively slots, as thick as the lode requires, running along the lode, and excavated as far up and down it as is feasible. The mining techniques are called overhand and underhand stoping, though there are numerous variations of them. In overhand stoping, the miners typically worked from platforms of stout logs wedged from side to side of the lode, while in underhand stoping they stood on the ore itself. After shot-holes had been drilled and blasting done, the broken ore was shovelled down wooden chutes to a lower tunnel for transport to the surface. A stope in hard granite can stay open for centuries, but huge falls of rock sometimes occurred, causing great damage to mine and men.

As the miners came on shift they had to get from the adit to the current working level, which meant a great deal of climbing steep ladders in addition to the grinding labour in the stope. Not surprisingly, heart disease was a contributor to the miner's short lifespan, as was silicosis, caused by working in fine dust, and which leads to wasting or consumption of the tissues. Another, but rare, affliction was ankylostomiasis, or hook worm, due to the unsanitary conditions underground. The worm lives in the intestinal tract and flourishes in faeces, causing a very unpleasant and debilitating illness which in one case infected 70 per cent of Dolcoath's workers. Hook worm is not indigenous to Britain, though it is found throughout the tropics and subtropics, so it may have come via a returned emigrant who had worked in, perhaps, Latin America.

The men worked by the light of a candle in the cap or in a lump of clay stuck to the wall, though acetylene lamps were eventually introduced. From the mid nineteenth century until proper miners' helmets came into use in the early twentieth century, workers in metal mines usually wore a 'tull': a linen skull cap under a felt hat which had been hardened by impregnation with resin and gave a measure of head protection. Waterproof clothing was eventually provided; hearing protection was unknown until modern times.

Ventilation might come from a through-draft from an adit to a shaft, but in headings it would be rather hit-and-miss. By the late nineteenth century proper ventilation driven by enormous electric fans became standard, but it was much too late for the thousands of men whose health was broken by the age of 40.

Women never worked underground in the mines of the southwest, as they did in some of the coalfields, but as a miner's wages in the mid nineteenth century were in the region of 10 shillings (50p) a week, they had to do their share by working as 'bal maidens'. This involved using hammers to break larger lumps of ore to smaller sizes as input to the next stages of processing, in which minerals were separated from gangue. This paid a shilling (5p) a day, or 6 pence (2.5p) per ton on piecework. Children as young as 8 worked on small tasks until they were old enough for the boys to graduate to the mine and the girls to bal work. Carefully posed photographs from the time show bal maidens in white pinafores and looking rather neat and tidy, but the daily reality must have been very different. After the hand crushing the next stages of mineral recovery (usually called mineral dressing) involved washing finely ground ore in water, causing the dense minerals to separate from the less dense gangue (processes called jigging and buddling). By the mid nineteenth century the whole system was overwhelmingly mechanized and concentrated in large dressing mills.

Miners can be somewhat superstitious, which is not surprising considering the dark conditions and the potential dangers of their work. In medieval times German miners had a whole pantheon of mythical creatures, such as gnomes who did no harm and mainly sat around looking busy but doing nothing, as in some modern gardens. The colourful costumes of garden gnomes are based on the clothing of sixteenth-century German miners. The Cornish equivalent was 'knockers', who could be heard in the distance. The rational explanation for this is obvious but, even into the late nineteenth century, some of the older miners still believed in them.

The coast of Cornwall has many small harbours, so smuggling was a useful source of extra income. Fowey, the main port of medieval Cornwall, had a very nice sideline in illicit goods. Many Cornish mines are on the coast and some miners were part-time fishermen and kept boats, so it is impossible to believe that they never indulged in a little contraband. That coast is also exceedingly dangerous for shipping, and wrecks were frequent, again providing a secondary income from cargo washed ashore.

Before we leave the south-west it is important to mention that its mineral riches were not limited to tin and copper. Tungsten was a significant by-product, and manganese occurs in the tin/copper veins. Manganese was, for instance, mined on a small scale at the delightfully named village of Doddiscombsleigh on the edge of Dartmoor. Another significant mineral was arsenopyrite, which is iron arsenic sulphide; arsenic hardens lead shot and printing type and is a powerful herbicide and insecticide. Other mines were mainly for iron and lead, the lead being rich in silver, though we'll come to lead and silver later in this book, and in the nineteenth century uranium from two Cornish mines was used by the Curies for research into radium, and for a short while in the 1940s it was used to feed Britain's nascent nuclear weapons programme. A related effect is that radon, a radioactive gas, is formed in granite areas and percolates into the atmosphere. Normally the concentration in air is negligible and we all breathe it, but the gas can accumulate in confined spaces such as mine workings, so radon-induced lung cancer may have been a cause of what in the nineteenth century was called phthisis, the then catch-all term for lung diseases.

Finally, we'll look at two smaller copper-mining areas. The mines at Coniston in the Lake District were high on the fells but, despite the harsh winter weather, they achieved notable production in the sixteenth and early seventeenth centuries. The smelting works at Keswick were destroyed in the Civil War. Conditions were, though, even worse in the copper mines of Snowdonia, which were worked, on and off, from pre-Christian times until the early twentieth century. The copper belt ran for some 15 miles from north of Porthmadoc towards Blaenau Ffestinog: about 36 mines have been identified. Most were very small operations employing only a few men, so they are little more than a footnote to Cornwall and Anglesey. Unfortunately for the miners, some mines were high on Snowdon where, in the practically Arctic winter weather, it was common for the men to have to dig through snowdrifts to get to the mine, and the ore was brought out on sledges or carried by men. Labourers were paid about 10 shillings (50p) a week in the mid nineteenth century; a sharp contrast with the circumstances of the Assheton-Smith family, who owned part of the land and had little concern for the copper workers as they drew £30,000 a year from Blaenau Ffestiniog's slate mines. It is hard to grasp such wealth; Lord Hertford had a house in Wales which he never visited, though the servants prepared dinner for 12 people every night in case he turned up. They were certainly 'on to a good thing', unlike the people who produced all the copper and tin for them.

4

Power to the Miner's Elbow

Mining takes a great deal of effort, usually in very difficult conditions: the valuable mineral has to be brought out—miners often talk about bringing it to grass—and moved to the processing area; waste rock has to be removed and disposed of; and the ever-present water, sometimes in copious quantities, has to be shifted from the workings. Overall, mines need a lot of power, and over the centuries some brilliant people have devoted great ingenuity to providing it, though with much trial and error. Good ideas always spread, so very similar machinery was used in most mining areas. We'll look at those common factors here and some special cases in the different industries.

It is, though, a harsh fact that the first flint mines in Britain were powered entirely by the efforts of people or animals in 2500 BC, and that very little changed until demand for materials such as lead to roof the medieval cathedrals and castles led to mining becoming a fairly large-scale industry around AD 1200, a span of almost 4,000 years. Men (and in coal mines, women and children) walked to and from their underground work and climbed primitive and dangerous ladders. Ore, coal, and gangue were carried out on human backs. Water was raised in buckets, and in wet mines half the working shift might be needed for that before mining work could start. Ventilation was virtually non-existent in many cases; where a mine had a horizontal adit, a vertical shaft might be sunk in the hope of getting a through draft, but even that was very imperfect.

The only 'mechanical' help was the jack roll, a very ancient, though still man-powered, device for hoisting ore or water to the surface instead of carrying it. This was a cylinder of wood, set horizontally, supported on two uprights and with a winding handle at each end: in essence, it was the windlass of a water well. A rope was wound round the cylinder so that one end was at the top of the mine shaft or pit and the other end at the bottom. With two men on the handles, buckets of water or baskets of ore could be raised, with the empty bucket going down for the next load, or taking something or someone down the mine. When the bucket reached the top of the shaft the jack roll men simply changed the direction of winding. On top of the fact that the jack roll winders worked in the open and in all weathers, this must have been cripplingly hard labour. The real drawback of the jack roll was its very limited capacity; at best only a few tons of material could be shifted in a day. Its advantages were that it was easy to build, labour was cheap, and it could be installed underground to reach deeper workings. Human labour and jack rolls could not cope with the increasing demand for metals and coal, but human ingenuity rose to the challenge, and far more powerful and effective machines evolved over the next few centuries. Despite its great antiquity, the occasional jack roll was still working in the Pennines in the twentieth century, wherever a few men got together to rework an old mine or a small new deposit.

Horse power, despite the considerable cost of buying and feeding the animals, could increase capacity and the horse gin, or whim, was invented in Germany in the sixteenth century, coming to Britain soon after via German miners imported for their hard-rock mining expertise. Figure 4 is a simplified version of what was involved. Two horses were harnessed to the opposite ends of the horizontal beam (only one is shown for clarity) and made to walk in a circle, guided by two lads. The rope is wound several times round the main drum, to give friction, and the ropes round the smaller drums raised and lowered mine output, water, etc., as required, with men at the top and bottom loading and unloading the buckets, or kibbles. When the kibble at the top had been emptied, the horses were made to walk in the opposite direction to repeat the cycle. There was usually a cover to protect the horses against the worst of the weather. Whims greatly increased capacity and were in use well into the nineteenth century. Steam engines had been invented by then, but there was the cost of buying and transporting coal, and the early engines needed a lot of mainten-ance, so the whim still had the edge in smaller operations. In any case, the horse dung would be a valuable by-product for the miners' vege-table gardens.

FIG 4 Horse whim.

To save the money spent on horses attempts were made to use windmills, but these were not reliable due to the vagaries of the wind, and attention turned to water power. While water in mines is the bane of the miner's life, as well as a risk to it from flooding, many mining areas have ample surface water, such as in the lead fields of the Pennines, so waterwheels came into extensive use. They had been used since ancient times for grinding corn and raising water, and a simple waterwheel could generate about three or four horsepower. That figure could be increased by building bigger wheels, and inventive people were soon building wheels 30 to 60 feet in diameter (the height of a six-storey building), so wheels could generate a lot of power. They were often boxed in to protect them from the weather: a high wind might blow the water off the wheel, or even damage the structure. There is an example of such a wheel at Killhope museum in Weardale, but the largest waterwheel in the British Isles is the Lady Isabella wheel at Laxey on the Isle of Man, which drained the local lead, copper, and zinc mine. It was built in 1854, named for the wife of the then governor, and is 72 feet in diameter. It is unusual in that the water came via a closed pipe from the reservoir, the pressure forcing the water up a sort of chimney to the top of the wheel.

Obviously, a waterwheel had to be supplied with water and insured against drought. Reservoirs were constructed to catch convenient streams, and ditches, or leats, were built to take the water to the mine; their remains often show up clearly in mining landscapes. A wooden trough, or launder, carried the water above the ground from the end of the leat to the wheel.

The leats were triumphs of the surveyor's art as the water had to hit the wheel at the right height, usually at its top, and had to flow in sufficient quantity to turn the wheel at the right speed, but neither so fast as to erode the sides of the leat nor so slowly that silting occurred. Leat men were employed to maintain the system. On top of that the leat had to connect with the spillway from the reservoir, which might be miles away over hilly country, and the builders wanted to avoid spending money on cuttings or tunnels. Although water will flow at gradients as small as 1 in 1,000, it was found that 1 in 400, about 0.14°, was a good compromise for these factors, but that often meant following the contours round the valley sides, with any side streams being fed into the leat. Even with modern laser-based equipment this would be a serious bit of surveying, and one can only admire their achievement. A leat might supply more than one wheel, so big stones would be inserted to split the water.

Waterwheels, reservoirs, and leats were big capital projects only justified by large-scale mining, and the costs required a wealthy backer or an organized company. The Duke of Devonshire, for instance, had large landholdings in Derbyshire and north Yorkshire: Chatsworth House and Bolton Abbey are still the family seats. In the nineteenth century the Duke financed water courses over a huge area of Grassington Moor and Greenhow, near Pateley Bridge, to power the mines which provided his income from mining royalties. If you drive from Pateley Bridge to Grassington, both of which were once mining towns, the remains of the old mines are spread along the roadside.

Waterwheels had many applications so a mine might have more than one wheel. The most obvious uses are for raising and lowering men, materials, and mine output, and for pumping out water. To connect the wheel to the hoist a large crank was attached to the wheel spindle and connected to a so-called flat-rod made of wrought iron or steel leading to a wheel at the top of the shaft. The advantage of this system was that a series of flat-rods, connected by iron brackets, could transmit power over considerable distances, in one case from Penzance to a mine on rocks 300 yards offshore, and similar distances were achieved at the Grassington Moor lead mines. The record is, perhaps, the Gawns mine on Bodmin Moor, where a 50-foot waterwheel drove flat-rods for a mile and a half.

The shaft wheel has two winding ropes wrapped round it in opposite directions; the same principle as the whim, but more powerful. When the kibble of ore reached the top, the winding wheel could be stopped by a simple clutch, and restarted in reverse when the kibble was ready to go back down. The kibbles were made of hazel wood and, when iron kibbles supplanted them, the price of hazel nuts in London fell sharply. Signals could be sent by bells connected to wires hanging down the shafts. This system worked but, as the empty kibble going down did not balance the weight of a full one coming up, the strain on one side of the waterwheel was heavy, so they had to be solidly built.

Like everything else in engineering, waterwheels had drawbacks. A big one was the reliability of the water supply: while reservoirs reduced the risk from drought, freezing in winter could not be prevented. The reservoirs themselves needed some maintenance to prevent catastrophic floods, and the leats had to be kept clean of rubbish and dead animals. The wheel needed attention when the wooden slats broke or the massive timbers of its construction rotted; later wheels were built of iron or steel. A major snag is that a waterwheel takes time to stop and restart and can't be reversed.

The waterwheel was expected to do a lot of work, and getting rid of water was important, initially by means of what was called the rag-and-chain pump. This consisted of an iron pipe hanging in the mine shaft. A continuous chain went down the pipe and back up on the other side of a wooden partition which protected the pipe from the balls dangling outside it. The chain was driven by the waterwheel and had iron balls attached at intervals which forced water up the pipe and dumped it at the top; Plate 7 is a sixteenth-century woodcut of such a pump. In practice, the iron balls damaged the pipe and leaked too much water back down the riser pipe, so they were often replaced by bundles of rags. If a waterwheel was, for some reason, not feasible, the pump could be powered by a jack roll, a horse, or even by men on a treadmill. In practice, the rag-and-chain pump was not up to the task of de-watering a very wet mine and better methods were needed, but we'll look at those later when we come to the steam engine.

While the rain that fed the streams was free, water was actually quite expensive because of the cost of the leats and could not be allowed simply to flow away downhill from the waterwheel. These prudent people solved that problem in all sorts of clever ways. One was to sell the used water from mine A to mine B, which was further down the valley. That involved building another leat to B, so some of the leats were eventually 10 miles or more in length. Mine B would be expected to pay some of the cost of the original leat to mine A, so you can imagine the disputes that arose about who paid for what and whether B was getting its fair share of the water, especially during a drought. Litigation could, as so often happens, be more expensive than the value being argued over. These issues could arise even when the same company owned both A and B, as the latter's manager would still have to show a profit, so the arguments between mines could be fierce. Violence and chicanery were not unknown.

Ideally, the mine management would want to use its own water as much as possible, so the water from the main waterwheel could be fed to a smaller wheel to drive the machinery used for crushing and concentrating the ore. It could also be fed down the shaft to drive another waterwheel, which pumped water and hoisted ore from mining levels below the main water level, which was an adit driven in from the hillside at a gentle slope to give access to the ore, and to drain the ground above it. But even the water coming down the shaft was made to work. It contained air, so the water was fed into a large, closed, tank at the shaft bottom. Water for the underground wheel was taken from the bottom of the tank, and air, at quite considerable pressure, built up at the top of the tank. From there it was fed through pipes to the working faces to give at least some ventilation. When compressed-air drills came into use in the 1880s, this air could supply only a few drills. Widespread use of these drills would eventually call for proper air compressors. When the water had been used as much as possible, it was allowed to drain out of the water level.

Waterwheels have been the queen of mining power sources. They are elegant pieces of engineering, powerful, versatile, and even graceful. There was little that a waterwheel could not do that steam and electricity could achieve; though, of course, those later sources were even more powerful and efficient than the waterwheel. It is, however, fair comment that without the production of metals and the accumulated wealth that the waterwheel made possible, steam and electricity could probably not have been developed.

The eighteenth century was a ferment of innovation and progress in mining, because when mines became ever deeper, the horse whim could not cope, and waterwheels, were vulnerable to frost and droughts. Attention turned, therefore, to the possibilities of using steam for raising coal and ore, and especially for pumping out the huge quantities of water that bedevilled Cornish mines and limited their output. This was a century in which Britain was regularly either fighting or financing wars, and as we have seen, copper was vital for sheathing the Royal Navy's wooden ships against the teredo worm which bored into hulls, and weed which reduced ships' speed.

As early as AD 100, Hero of Alexandria had built an engine that turned a copper ball mounted on a spindle by blowing steam from two nozzles. This was no more than a curiosity, and while Arab scientists may have experimented with steam, it was not until the late seventeenth century that progress was made. In 1662 Robert Boyle published his famous law to the effect that heating a fixed volume of gas raises its pressure. This idea circulated rapidly throughout educated circles in Europe, and designs for steam engines started to emerge. The first practical one was that of Thomas Savery (*c.*1650–1715), who developed a pump that could raise water through a height of about 50 feet, which might have been adequate to supply a palace but was not much use in a mine. Savery had, though, patented his pump so very skilfully that Thomas Newcomen (1663–1729), who developed the first effective steam engine, was forced into partnership with him. But before we explain how Newcomen's engine worked, let's look at what it did.

Coping with the continuous flows of water in a mine, or 'de-watering' one that had been flooded, were vital necessities. The Newcomen engine made those tasks feasible because it could drive a pump rod up and down several times a minute and several feet each time, which allowed water to be lifted by that distance and tipped into a cistern to feed the next lift. The pump rods were substantial timbers up to 40 feet long and one or even two feet square, a size that was needed to carry the weight of all the rods in a shaft 1,000 feet or more deep. They were imported from the Baltic, and then had to be taken by horse and cart over the Pennines or across Cornwall to where they were needed. Timber was even occasionally imported from British Columbia in western Canada, though the logistics of that in the eighteenth century beggar belief.

The early engines used bucket pumps to do the lifting. Bucket pumps were essentially just cylindrical iron tubs with holes in the bottom and flaps to cover the holes. When the pump rod went down, the bucket was dipped into the sump at the shaft bottom, or into a higher cistern, and was filled with water, and when the rod went up the flaps closed. At the top of the lift the water was tipped into a cistern and the cycle repeated. This was much more effective than the rag-and-chain pump, but was still not very efficient as there was a lot of spillage. The real breakthrough came with the invention of the plunger pump in which a series of plungers forced water up a pipe from one cistern to the next, with simple valves to stop the water flowing back down, though a bucket pump had to be used to get water from the sump at the bottom of the shaft. This system could cope with very large amounts of water and made it possible to work big, deep, wet mines.

The working of Newcomen's engine is shown in outline in Figure 5, though many details are omitted. To enable the rods to move down, and the piston in the cylinder to move up, steam is allowed to flow in under the piston. The steam was at a little more than atmospheric pressure, so it counteracted the atmospheric pressure trying to keep the piston down. As the piston slid up the cylinder, the pump rods went down the shaft by their own weight, forcing the plungers and raising the water. When the piston reached the top, the steam valve was closed and the condensate valve was opened, spraying water into the cylinder, rapidly condensing the steam, and producing a nearly perfect vacuum in the cylinder. Atmospheric pressure then drove the piston back down and the cycle could restart. Since it was atmospheric pressure that did the work, the first steam engines were called 'atmospheric engines'. The piston and pump rods were connected with strong chains to the ends of the rocking beam, which meant that the two rods were always vertical, minimizing wear. The water on top of the cylinder sealed and lubricated the piston.

A major problem with Newcomen's engine was that atmospheric pressure on a piston 60 inches in diameter is about 18 tons, but the pump rods and water being pumped weigh very much more than

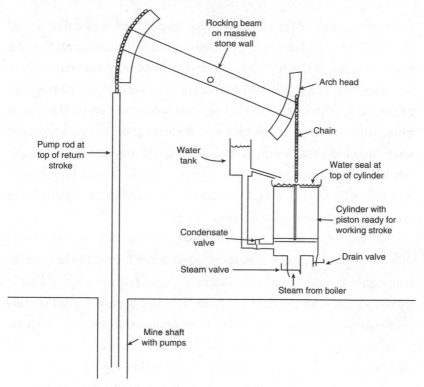

FIG 5 Principles of Newcomen's 'atmospheric' engine.

that, which would make the engine incapable of doing anything useful. The solution was to balance the weight in the shaft using a 'balance bob' so that the engine only had to provide power to lift the water. The bob was a heavy weight, or box full of rocks and scrap iron, attached to the pump rod via a long pivoted lever. As the rods went down, the box was raised on its lever, and vice versa when the rods went up, thus creating a see-saw effect. The bob would be mounted on the surface near the shaft, and in deep shafts extra balance bobs were installed in chambers cut into the rock adjacent to the shaft wall.

The water tank at the top was fed by a pump that was linked to the rocking beam, so water supply was automatic, but, ironically, in the first Newcomen engines the steam, condensate, and drain valves were operated by a boy. With the engine going at 10 strokes per minute, that was quite a challenge, and it was not long before the valves were linked to the beam so that they, too, worked automatically. It has been alleged that a boy called Humphrey Potter first did this with ropes. That seems very unlikely, as the engine man would probably have taken his belt to a boy who fiddled with the machine, and young Potter might well have regretted his initiative.

The first Newcomen engine was installed at a colliery in Staffordshire in 1712 and was a great success, but a serious disadvantage was its coal consumption. This was because at the bottom of the piston stroke, as seen in Figure 5, the steam has just been condensed with cold water and the cylinder is cold. The first inflow of steam does no more than heat the cylinder, which is then cooled again on its working, atmospheric, stroke. This waste of heat did not matter at a coal mine, as small coal, which could not legally be sold, could be used in the engine, but it was a major drawback in Cornwall given the costs of the coal and transporting it by sea from South Wales or Somerset and overland to the mines, and a duty which was charged on all seaborne coal, even if it was shipped within the country. So, while a few engines had been erected in Cornwall by 1722, the mines mainly went back to using waterwheels until the duty was abolished in 1739. Thereafter there was a spate of engine-building, and within three years about 25 were in operation. Generally, the building of an engine and its house was well within the abilities of the local smith, carpenter, and stonemason, with the exception of making the boiler. At first the latter came from Staffordshire, and were initially made of brass, then eventually of iron when foundry techniques improved. In due course, the mining districts throughout the country developed foundries and engine works of their own, sometimes employing thousands of skilled artisans, and numerous engineers emerged up and down the land.

Where it was not feasible to mount the beam directly over the shaft, flat-rods could be used to connect to a remote shaft. Figure 6 shows the idea: a rod coming from the waterwheel's crank or the steam engine was connected to a triangular 'angle bob' at the top of the shaft, the rocking of which made the beam in the shaft move up and down to drive the plunger pump. It sometimes happened that the vertical shaft would find a rich vein of ore and the shaft would be driven at an angle to follow the quick money, but a second angle bob

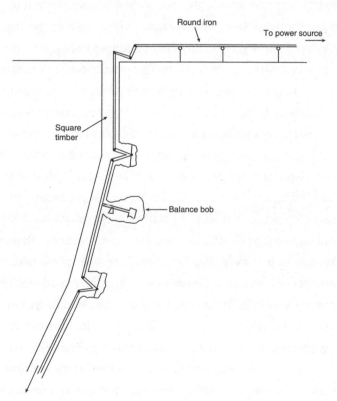

FIG 6 Flat rods, pump rods, and balance bobs.

could cope with that. Handling all this heavy timber in the shaft was a specialist job done by 'pit men', while the timber was called 'pit work'.

The Newcomen engine was improved by John Smeaton (1724–92), who achieved greater power by building one with a cylinder 72 inches in diameter—the usual way of describing an engine is by its bore and the length of its piston, the stroke, which was commonly in the region of 8 or 10 feet—but the real breakthrough came in 1775 from the genius of James Watt (1736–1819). Watt saw that the fundamental weakness of the Newcomen engine was the alternate heating and cooling of the cylinder, several times a minute at full speed, which wasted about 80 per cent of the steam. His solution in 1775, shown in outline in Figure 7, was to feed the steam not only to the bottom of the piston but also to its top, where it also supplied a steam jacket ensuring that the cylinder was always hot.

The upward stroke of the piston is, as with Newcomen's engine, done by the weight of the pump rods in the shaft. The engine had three valves, A, B, and C, all cleverly operated by rods linked to the movements of the beam, so that the valves opened and closed at just the right times in the engine's cycle. In Figure 7, only valve B is open, allowing the steam in the top of the cylinder to flow through the transfer pipe into the bottom of the cylinder as the piston moves up. When the piston reaches the top of the cylinder and the downward, working, stroke starts, valve B closes and A and C open. Steam now rushes in via A to drive the piston down, while C allows the exhausted steam from the previous working stroke to flow into the condenser, aided by the vacuum pump. The Watt engine was no longer atmospheric, as the work was done by steam, not air; but since the steam was at low pressure, balance bobs were still needed. To keep air out, the cylinder had a sealing gland, or stuffing box, packed with hemp and tallow. The piston was no longer lubricated with water, so hemp and tallow were used until proper piston rings were invented. Eventually, the massive

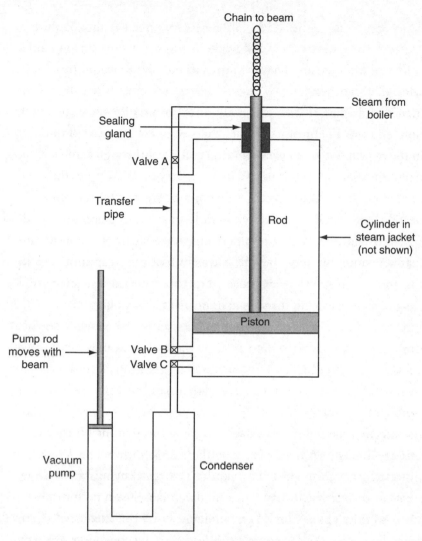

Chain to beam

Steam from
boiler

Sealing
gland

Valve A ☒

Transfer
pipe

Rod

Cylinder in
steam jacket
(not shown)

Piston

Pump rod
moves with
beam

Valve B ☒
Valve C ☒

Vacuum
pump

Condenser

FIG 7 Principle of the Watt engine.

wooden rocking beam was replaced by elliptical cast-iron beams, which were often of rather elegant design, even beautiful.

The main advantage of Watt's engine was the saving of heat, on top of which it ran more smoothly than Newcomen's. In order to turn the fruits of his genius into a commercial reality Watt went into partnership with Matthew Boulton, a Birmingham manufacturer of metal items. Their business model, as we might now call it, was that they neither sold engines nor would they allow anyone else to erect a Watt engine. Instead, they built and installed the engine and charged the mine an annual fee based on the saving of coal over a Newcomen machine. That was typically in the region of £2,000 a year or more, so Boulton and Watt soon became very rich.

Both Newcomen's and Watt's engines were single-acting in the sense that the power stroke was only once in a cycle, and it was a natural improvement to develop the double-acting engine, with both strokes doing work. However, a chain obviously cannot be pushed, so the piston rod was connected directly to the beam, but the rocking of the beam would also make the piston rod rock back and forth in the cylinder, increasing wear and friction. Watt's genius was equal to that, and he devised a way of connecting the rod to the beam so that the piston rod, and the pump rod in the shaft, would always be parallel and vertical. Watt regarded this 'parallel motion' as his most brilliant innovation.

Another name to conjure with is Richard Trevithick (1771–1833), a Cornish man who, despite having been to school, was scarcely literate. He did, however, have an intuitive gift for solving engineering problems which made him the engineer to several mines when he was only 19 years old. Watt's engine still consumed a lot of coal, and Trevithick saw that greater efficiency could be achieved by using steam at a pressure of about 40 pounds per square inch, which the cautious Watt thought to be too dangerous. Trevithick's engines were very successful partly because, for a given power, they were much smaller than Watt's, and were easily moved. In 1801 Trevithick used one to build a steam-driven carriage which laid the foundation for the eventual development of steam traction engines and railway locomotives. For engines operating at much higher pressures than 40 psi and with heavy steam consumption, such as those driving factory machinery, iron foundries, and high-speed winding engines for mines, Trevithick designed the Cornish boiler, which had one large tube down the middle for the fire. Throughout Britain his engines were known as 'Cornish' engines. Trevithick led a very varied life, including working in South America, but died a pauper. He is buried in Dartford, Kent, though the site of the grave is not known.

The success of Watt, Trevithick, and other engineers is shown by the fact that, in the boom years of the 1850s, some 650 beam engines were working in Cornwall, plus another 60 or so in Devon, which represents a colossal output from Cornish foundries and engine works such as Perran Foundry and many others. Sadly, the decline of the industry meant that by the 1920s only about 20 engines were still in steam.

A further evolution of the steam engine came from the realization that high-pressure steam was not used effectively in a single cylinder, as a good deal of pressure was still left at the end of the stroke. Thus, in 1781, Jonathan Hornblower (1753–1815) patented the compound or double-expansion engine, which had two cylinders working the same beam. One was 19 inches in diameter and used high-pressure steam, while the other 24 inches in diameter to exploit the residual steam—a far cry from the cylinders of as much as 100 inches needed with low-pressure steam. The final stage was small, powerful engines, mounted horizontally and driving wheels for crushing mine output, hoisting and pumping, driving factory machines, and, of course, for locomotives and steamships, leading eventually to triple-expansion engines and ultimately to steam turbines.

In the late nineteenth and early twentieth centuries electricity began to replace steam power and waterwheels. The process was initially slow as many mines had to rely on their own generators, but with the advent of the National Grid in 1936, electricity steadily came into universal use, though techniques had to be developed to prevent electrical sparks causing explosions in coal mines. However, even in the 1960s many coal mines still had steam winding engines (which were magnificent to watch when running at full power). To supply the necessary volume of steam, they used the two-tube Lancashire boiler; people working on the surface made a point of keeping in with the boiler man, as there was a guaranteed supply of hot water for tea and usually a small stove on which bacon could be fried.

As steam gave way to electricity the steam engines were demolished and sold for scrap, leaving, in Cornwall, the graceful but evocative ruins of the engine houses. Happily, groups of skilled enthusiasts have preserved some of these wonderful engines, which can be seen up and down the country. Some are still in steam, but the costs of testing the boilers for safety are high, so others are (ironically) run by electricity. Whatever the power source, they are a splendid sight, with gleaming brass, and complex gears and control rods moving in harmony with the engine.

Another important use of steam power, particularly in Cornish mines, was for a man-riding engine. We have to imagine the miner walking perhaps a couple of miles to the mine and then descending a near-vertical ladder to the working level: a depth of 1,200 feet was not unusual. The man then had to walk down narrow, often twisting, tunnels to the face, do his work, and finally retrace his steps up those 1,200 feet of ladders. To put that in context, Blackpool Tower is 520 feet high, the Eiffel is just under 1,000 feet, and convicts in the mid nineteenth century might be sentenced to three months' hard labour on the treadmill, but prison regulations limited the day's task to not more than the equivalent of climbing 1,200 feet. However hard the punishment, the convict's sentence was for a limited period, but the miner was doing much more than twice the convict's labour, and was doing it as his life's work. Small wonder that miners' lives were short.

The man-riding engine used a rod in the shaft, separate from the pump rod, and with wooden steps about 2 feet square attached to the beam at intervals of about 12 feet, with corresponding steps on the shaft wall. The miner simply stood on the beam step until it had risen to the level of the shaft step and, in the short interval when the beam was going neither up nor down, changed places with a man standing in the step in the shaft wall, which had to be done quite briskly. In such a way he could ascend or descend 1,200 feet in about 25 minutes. Unhappily, the steps became slippery with muck from the mine and grease from the iron plates which connected the lengths of timber of the shaft beam, so accidents occasionally happened, though these were rare. The exception that was the Levant mine disaster, which we will encounter in Chapter 11.

The work of breaking rock or coal was, from 2500 BC until the time that gunpowder came into use in the sixteenth century, done with a hand pick or plug and feathers. In that method, a hammer-and-chisel-like borer was used to drill a hole in the face into which two iron slips—the feathers—were jammed. The plug—an iron wedge—was hammered in between the feathers, breaking off more substantial lumps of ore. In metal mines, fire-setting was sometimes used. The use of gunpowder required holes to be driven with a boring bar and heavy hammer, the shot was fired, and the broken ore had to be shifted by hand. This was some improvement, though the fumes from black powder in an ill-ventilated mine meant very bad working conditions and limited the miner's life span to about 45 years in all too many cases. Dynamite, invented in 1866, was more powerful and its fumes marginally less hazardous, but the real advance in rock-breaking was the invention of the pneumatic rock drill and its exploitation by the firm of Holman Brothers in Camborne.

John and James Holman took over their father's boiler-making business in 1881 and soon went into partnership with James McCulloch, who had invented a drill driven by compressed air. This proved to be a success, not only in Cornwall but also in South Africa in the rapidly developing gold mines of the Witwatersrand. The original drill created very fine dust which was bad to work in. The solution was a jet of water passed down a hollow drill rod to clear the hole, which cured the dust problem but involved wet work for the driller. Nowadays, miners have waterproof clothing and hearing protectors, as the drills are very noisy in the confined conditions of a mine.

Finally, we'll look at machines that took over the back-breaking work of loading ore or coal into small trucks, called tubs, or onto a conveyor belt. There are several types of that sort of machine, and we'll look at the specialized coal-loaders in Chapter 6 and at some very advanced systems used in gypsum mines in Chapter 7, so for now we'll illustrate the principle of ore-loaders with the Eimco over-shot loader. This was originally developed in the 1930s by the

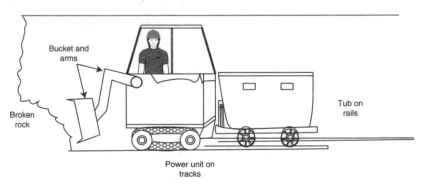

FIG 8 Principle of an Eimco loader.

splendidly named Eastern Iron and Metal Company of Salt Lake City, Utah, but is now in very widespread use worldwide and extensively improved from the first version.

In the highly simplified diagram in Figure 8, the unit is powered by electricity via a heavily armoured cable, compressed air, or a small diesel engine. It is mounted on tracks or wheels and can crowd its bucket into the rock pile. The operator stands on a platform at one side and, when the bucket is full, operates the arms, which are more sophisticated than the simple L-shape in the diagram, so as to throw the bucket contents over the machine into a tub, a powered mine car, or onto a conveyor. With track-mounting this type of machine can dig its way up or down a slope and can clear to the sides of the tunnel.

Overall, the engineering history of British mining is an impressive story of ingenuity, determination, and business acumen that, over the past few centuries, has enabled mines to work on a large scale, and at considerable depths. It has allowed long tunnels to be driven and vast amounts of water to be removed from mines. Without the achievements of remarkable people such as James Watt, Richard Trevithick, and countless others whose names do not come down to us, it is fair to say that our way of life would not be as it is. Perhaps most of all they made the life of the miner less laborious.

5

From Castle Roofs to Organ Pipes: The Lead and Zinc Mines

Lead and zinc are two very versatile metals. Lead is used in batteries, for roofing, and much else, while one of zinc's important applications is galvanizing steel to make it rust-proof. Conveniently, lead and zinc minerals occur in the same ore body. Barytes (barium sulphate) and fluorspar (calcium fluoride) can also occur in conjunction with the lead and zinc minerals, but until their many uses, which are described at the end of this chapter, were discovered they were treated as gangue. A valuable aspect of lead mining is that the main mineral, galena (lead sulphide), often contains silver, which for hundreds of years meant that some lead mines were really silver mines (silver and gold are covered in Chapter 10). Lead and zinc mining have a remarkable history, with some notable feats of mining engineering, together with the great skill and dogged endurance shown by the miners and their families. The story involves both mine work and farming by the families, the vibrancy of their vanished communities, and the evidence of their work in the landscape.

The Lead and Zinc Mines

In the mines the lead and zinc minerals typically occur in steeply inclined shoots, or veins, many feet high and running for long distances. Some veins are quite thick but others are only a few inches wide, though there are usually numerous veins in a mine. Another type of vein is the flat, called a 'pipe' in Derbyshire, which is more or less horizontal, often a few feet thick, and can be extensive. Zinc minerals, barytes, and fluorspar were seen as waste or gangue until the eighteenth century, so we'll cover their history after that of lead.

The Romans exploited lead and silver from the Mendips as far north as the northern Pennines, the main evidence being lead ingots inscribed with their origin. The most striking use was the lead lining of the baths at Bath. The workforce may have included slaves, convicts, and prisoners from local tribes, and there are suggestions that some of these poor unfortunates even spent most of their short and miserable working lives underground, but there were also native miners. Lead and zinc occur in many parts of England, such as the Mendips, Shropshire, and the south-west; in Wales in Cardiganshire, Montgomeryshire, and Flintshire; and in Scotland. The major sources lie in central and northern England, where the lead/zinc fields start around Matlock in Derbyshire and, as Figure 9 shows, run for about 130 miles to Weardale in Cumbria. Flintshire was second in importance to the Pennines.

In mid Wales, mining was a major source of employment as an alternative to poor agriculture. There were many small mines, one of which was the Van mine near Llanidloes which was worked on a small scale from Tudor times, though it is probably older than that. It later grew to be an important mine, and in total produced about 100,000 tons of lead ore, 30,000 tons of zinc minerals, and no less than 770,000 ounces of silver. The workings were in massive flats, and there are still large reserves deep underground. The mines closed in the 1950s in the face of foreign competition, so it is questionable

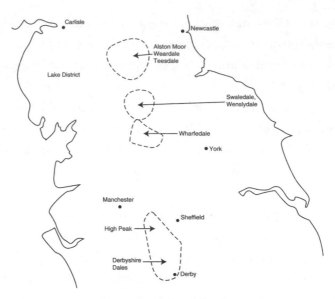

FIG 9 The Pennine mining fields

whether those reserves can ever be exploited, but we'll look at the future of British mining in the final chapter.

The Flintshire mining area was large; an aerial survey revealed signs of 250 mines. Roman remains in the area suggest that they worked the lead field, and there is evidence of medieval activity. For 100 years from the late seventeenth century, the London-based Quaker Company developed mining in Flintshire, sinking shafts and finding rich veins. The deposits were worked intensively throughout the eighteenth and nineteenth centuries, to such an extent that in 1850, 11,500 tons of metallic lead were produced, some 12 per cent of British output. Mining finally ended in the 1960s, after practically 2,000 years, but such longevity has been common in British mining. The landscape has signs of dressing floors (where galena was separated from gangue), tips, horse whim circles, and miners' cottages, some of which have been converted to modern houses. Halkyn mine near the town of Flint was notable, as its workings were in Carboniferous limestone under the Flintshire coalfield. This gave easy access to coal for steam engines and smelting, but Halkyn also mined vast quantities of limestone in addition to lead.

The Scottish mines were relatively small, compared to the Pennines, and were at Leadhills and Wanlockhead in Dumfries, and at Strontian in Argyllshire. Wanlockhead had four mines which worked from about 1675, New Glencrieff mine being one of the richest lead mines in Britain. It closed in 1931, but an attempt was made to reopen it in the 1950s, but it failed due to high costs and the then low price of lead.

The Strontian mines were worked for some 200 years from the beginning of the eighteenth century. They are significant for the discovery in 1790 by Adair Crawford of strontianite (strontium carbonate), who named the mineral for its location. Strontium is chemically similar to calcium and is used for red colouration in fireworks and in the extraction of sugar from sugar beets. The mines had a revival in the 1980s, producing barytes as a heavy filler for drilling the North Sea oil wells, but the open-cast operation destroyed much of the evidence of the earlier workings.

We'll mainly look at the very rich Pennine mines. The record of post-Roman mining is scanty but becomes a little clearer in Saxon times when lead is recorded as being sent from Wirksworth in Derbyshire to Ely in the ninth century. In 1086 the Domesday Book mentions that 250 pounds of Derbyshire lead had been due to King Edward the Confessor, which at the usual level of royalty implies annual output of a ton or so, but that may be an under-estimate. Between the twelfth and fifteenth centuries demand was considerable to supply lead for roofs and water pipes in castles and cathedrals. Lead was exported to Germany via the River Humber and to London from Yarm on the Tees. The cobbles in Yarm's High Street are ballast from medieval ships that came to trade for wool and lead: they left the stones, and the metal was used to ballast the wool on the outward voyage. There was also extensive inland trade; in 1189 100 cartloads of lead were sent from Derbyshire to Waltham Abbey in Essex, a distance of more than 100 miles as the crow flies. There was money in lead, and the merchants of Newcastle and York became rich and powerful on the lead and wool trades. The sixteenth-century Hall of the Merchant Adventurers in York, which includes a hospital, is testimony to their wealth and philanthropy. Others became rich on lead: for instance, it was mined by the monks of Fountains Abbey, and in the twelfth century King Stephen granted mineral rights to the Bishop of Durham, not only to pay for the cathedral but also to maintain troops against Scottish raids. The Rector of Stanhope was entitled to 10 per cent of the Durham royalties, which produced £12,000 a year in the mid nineteenth century, by far the richest Anglican living.

The Derbyshire miners were very skilled and knowledgeable; so much so that, when a major discovery of silver/lead was made at Coombe Martin in Devonshire in 1294, miners from Derbyshire were sent there to impart the necessary skills. One wonders what they made of going somewhere they had probably never heard of, and how they coped with the journey and dialects that were practically mutually incomprehensible when they arrived. There would almost certainly have been no going back. The Devonshire mines prospered for 200 years, and Derbyshire men were similarly sent to open new mines at Grassington and Appletreewick in Wharfedale, which are now delightful tourist spots but were once dirty mining villages. As the industry grew, laws and customs for the mines were developed from about 1100.

PLATE 1 A small part of central Cornwall in the heyday of mining.

PLATE 2 The legacy of mining—Coppermines Valley near Coniston in the Lake District.
The white building is now a youth hostel but was once part of the mines.
The cottages were miner's housing with gardens.

PLATE 3 Botallack Mine, Cornwall, an amalgamation of many small mines.

PLATE 4 A buddle used to concentrate cassiterite

PLATE 5 Jimmy Rowe overhand stoping at Geevor in Cornwall, 1920s

PLATE 6 Parys Mountain in 1785, painted by Julius Caesar Ibbetson. Two men work a jack roll while others break lumps of ore. This was done in practically all weathers.

PLATE 7 A ball-and-chain pump from a sixteenth century German woodcut.

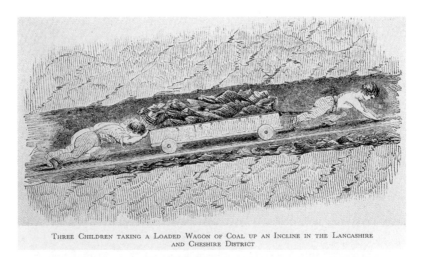

THREE CHILDREN TAKING A LOADED WAGON OF COAL UP AN INCLINE IN THE LANCASHIRE
AND CHESHIRE DISTRICT

PLATE 8 Boys hauling coal, from the Report of the 1842 Commissioners.

PLATE 9 The restored Laxey waterwheel, Isle of Man, the biggest wheel in the British Isles, 72 feet diameter and a masterpiece of design.

PLATE 10 The crowded mine shop at a Pennine lead mine. The facial expressions say much about their lives.

PLATE 11 The man-engine at Dolcauth. A posed flash photo, in reality the men used acetylene lamps.

PLATE 12 Bickershaw colliery at Leigh in Lancashire, a typical large coal mine with four shafts accessing different seams. Opened in 1877, it employed 3,000 men. It was closed in 1992, though millions of tons of coal remained.

PLATE 14 Penrhyn slate quarry, showing the vast amount of work done by quarrymen.

PLATE 13 Testing the roof at Llechwedd slate mine. The ladder is 60 feet high and the task was done by candlelight.

The Lead and Zinc Mines

In the same spirit as the Stannaries of Cornwall, the customs of the lead-mining districts in the Mendips were formalized under Edward IV (1442–83). In Cumberland the lead miners of Alston Moor were given legal privileges from the thirteenth century and, by the reign of Henry V (1386–1422), a court of mines had come into being, the king's officers having no authority to serve writs. Similarly, Derbyshire was divided into many small 'liberties' for the purposes of lead mining. However, laws or not, there were many disputes, sometimes very violent, over boundaries and the rights to water and timber.

The lead industry underwent many changes, but independent miners dominated the mines until about the mid-eighteenth century. They would contract with the mineral owner for the right to mine a 'meer'—about 30 yards—along the vein. However, when the mineral owners wanted easier administration and raising of capital for opening and drainage of larger mines, the meer system was abandoned in favour of areas or sets to be mined, and the free miners became employees. The outcome was that wealthy men, and families such as the Blacketts, acquired mines such as those at Allendale in 1694 which the family ran for the next 200 years. The London Lead Company—the Quaker Company—was formed in 1692 and continued in business in northern England until the late nineteenth century. The Duke of Devonshire had large land holdings in Yorkshire and financed the Grassington Moor mines in Wharfedale. He was also heavily involved in Derbyshire.

Mining techniques evolved slowly, but for hundreds of years the basic miner's tool was a pick, or gad, with a short shaft for use in confined spaces, with which he picked at the ore until a piece broke off. The next evolution was to use a heavy hammer, a maul, to drive the pick or gad deeper into the ore face, breaking off larger lumps. Plug and feathers broke off more substantial lumps of ore. Fire-setting was common from pre-Roman times but became less so after the seventeenth century. A fire was built against the face causing the rock to peel or blow off. Sometimes water was thrown onto the face to shatter the ore. It was highly skilled but sometimes dangerous work. Because of the smoke and fumes, fire-setting could only be used with the agreement of the men in other sections of the mine or at specific times.

The very earliest miners simply worked the ore they could see on the surface, making a narrow trench that meandered to follow the vein and which could be up to 50 or 60 feet deep. To go deeper, pits—shallow shafts—were used, and the miner took whatever ore he could reach, sometimes tipping the waste from each pit into its predecessor. This method was used as late as the eighteenth century in fringe areas. Another mining technique was hushing, in which a dam is built on the hillside and a torrent of water released to wash away the surface, leaving the miners to pick out any exposed ore. If the vein looked rich, they would loosen the surface and the hush would be repeated, with an iron grid or cross-trench to catch the heavier galena frag-ments. Some of the hushed areas, such as those near Yorkshire's Gunnerside Gill, are huge, as this process could be repeated over many years, but often the men worked in atrocious conditions.

The main limitation to underground mining was that the bouse, as ore was called, had to be raised in buckets by a jack roll. The weight of the rope, added to the bucket of bouse, limited small-scale mine shafts to about 50 feet deep, though larger mines used a single haul for as much as 200 or 300 hundred feet. As the mine got deeper a short tunnel, or level, was commonly driven for a short distance and a new, completely underground shaft, or winze, with its own jack roll was sunk; this process could be continued three or four times down to several hundred feet. It was very inefficient, as the miner at the bottom needed someone to carry the ore to the foot of the next shaft, each jack roll needed one or two men, and that was multiplied until the ore reached the surface. The miners climbed to work on primitive ladders set into the shaft wall, or by stepping into holes left in the stonework where a shaft was lined in soft ground.

The ever-present water was, for centuries, carried in buckets and hauled out by the jack roll, and it was not until about the fifteenth century that drainage adits, called soughs in Derbyshire, started to be used. They were driven from a nearby valley but that required considerable capital, knowledge of where the veins were, and surveying skill. The water adit allowed deeper work as water had only to be raised to drainage level, and not to the day or grass, and where the adit reached a shaft, much better ventilation was achieved. The adits were lined with stone arches in soft ground, which was work of great skill. Some adits were very long: the Great Level driven from Alston in Cumbria to drain the Nenthead mines was eventually about 5.5 miles long and took 50 years to drive, but found little ore. Driving adits was slow because, before powered drills came into use, a tunnelling team would be lucky to advance even as much as 20 feet in a week of day and night working in rock as soft as limestone. Sometimes an adit served several mines, which agreed to pay proportionate costs, but that led to many disputes. By the eighteenth century many of the adits were made wide enough for horse-drawn wagons running on oak, and eventually iron, rails. The horses were protected from the water by leather coats, but the men and boys who drove the horses were more easily replaced and so not protected.

Most veins in lead and zinc mines are at steep angles, so in the early trenches stopes were worked downwards. When the trench became so deep that the miner could no longer throw the ore to the surface, a layer of heavy timber was inserted, the miner threw to that, and a mate shovelled it to the top, a system that could be repeated a few times. In the underground mine stoping could be worked downwards (underhand) or upwards (overhand), with the ore being thrown down a shaft or into a wooden chute leading down to the adit. Lighting was usually by candle, sometimes replaced with acetylene lamps in the late nineteenth century, and ventilation was hit and miss; sometimes the candle almost faded out. As stoping proceeded strong timbers were inserted across the worked-out vein and the deads, or waste rock containing no ore, were stacked on the timber to make a working floor over the level below. Each level had such a timber roof, and the weight of deads kept the walls stable. Excess deads were taken to the surface and dumped; these dumps can often be clearly seen and are part of lead mining's landscape legacy. When the timber could take no more weight a new set of timbers was put in. This process could be taken to considerable heights. Not all veins were steep: for example, at the Boltsburn mine in Weardale, flats ran for 2 miles and, exceptionally, were up to 20 feet thick; this made Boltsburn the richest mine in the north for 40 years from the 1890s.

It was not until the late seventeenth and early eighteenth centuries that proper shafts were routinely sunk to depths of hundreds of feet, as that needed much capital which could only come from organizations such as, among many, the London Lead Company and the Blackett-Beaumonts. Deep shafts also needed more powerful methods for winding than the man-powered jack roll, and this came from the horse-gin and the waterwheel. Steam engines were used quite early in Derbyshire for pumping, but winding came only in the late eighteenth century. They were slower coming into use further north as coal there was more expensive, but by the early mid nineteenth century steam was widely used.

The Lead and Zinc Mines

The first explosive was black powder, later complemented by and then supplanted by dynamite. Shot holes were drilled with a hammer and boring bar, and from the 1880s by pneumatic drills. These were slow to come into use at first, partly because of innate conservatism but also, no doubt, fear that the greater output would lead to loss of jobs. The powered drills produced more, and finer, dust and made working conditions even worse in poorly ventilated mines where silica was common in the ores. That was eventually ameliorated by pumping water down a narrow hole in the drill and bit, but that made working conditions even wetter.

The bouse from the mine was taken to bouse teems for processing. 'Teem' comes from an Old Norse word meaning 'to pour', and meant the large stone bin into which the bouse was poured. The bouse was taken out, sluiced with water if sludgy, and smashed to small sizes with hammers and a bucker, which was a heavy iron plate fastened to a wooden handle, somewhat like a heavy carpet beater. The crushed bouse was then riddled into coarse and fine fractions, the coarse then being 'sieved' in a vat of water to separate the heavy mineral from the less dense gangue. Initially this was done in a hand sieve and a skilled washer could separate three layers: light gangue on top to be discarded; galena on the bottom; and a middle layer of mixed gangue and galena which would be crushed and washed again. The sieve evolved into a semi-mechanized hotching or brake tub, with a long lever to move a much larger sieve up and down. The finer fraction was sent to the buddle, where a slurry was poured down a slightly sloping surface, allowing heavier particles to separate near the head. All these processes were mechanized in the latter part of the nineteenth century, from crushing to the final refinement in the powered shaking table, which was a large metal tray, sloping gently, shaking rhythmically, and with metal bars across it to catch the heavier minerals enabling the lead and zinc ores to be separated; zinc sulphide, sphalerite, is less dense than galena.

Until machinery was introduced, bucking and bouse washing were mainly done by women, old men, and boys from the age of 10 or 11, often working longer shifts than the miners. They usually worked in the open in all weathers, unless the water supply actually froze or there was a drought. Payment was based on output, so in bad weather they had no income. When machinery came into use it was much more valuable than people, so covered sheds were built. The downside was that using machinery was men's work, so the women lost the income they had contributed to the household's meagre budget. Large mills required a lot of capital and tended to be at the major mines, or central to a group of mines. The noise in a mechanized mill is indescribable.

To summarize all this, Figure 10 is a simplified sketch of Killhope mine in the northern Pennines, which worked from 1853 to 1916. The diagram shows the early small pits and the large hushes on the fell top, the main adit, or horse level, and cross-cut levels into the veins of ore. The dotted lines show how the mine was developed in depth. Park Level Mill used powered crushing machinery. The buildings near the mine entrance were offices, workshops, and the mine hostel or 'barracks' that I'll describe later. Killhope is typical of many mines where the valleys were steep-sided and an adit could be driven to gain access to large amounts of ore. In Wharfedale, on the other hand, the hills roll more gently, so above Grassington and at Greenhow near Pateley Bridge the mines were shallower but numerous: their remains can be seen at the roadside. Power came from waterwheels, of which 40 once worked in Wharfedale, transmitting their output as far as 600 yards to individual mines using flat-rods—remarkable engineering. Greenhow benefited when

The Lead and Zinc Mines

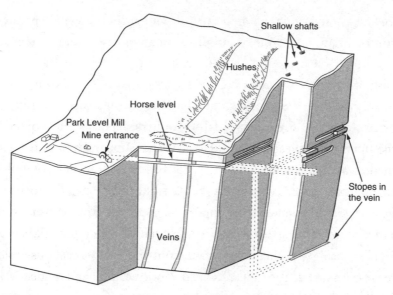

FIG 10 Sketch of Killhope Mine

Bradford Council drove a tunnel in the mid-nineteenth century to provide water for the city, and that had the added benefits of draining a large area for mining and revealing several rich lead veins.

The details of the refining of galena into lead or sphalerite into zinc are a little outside our scope here, but I'll indicate the main processes. The medieval method used the bole, in effect a large bonfire placed in a windy location with a burden of several tons of ore and at least as much wood for fuel. After 1580 this was rapidly superseded by the ore hearth in which a fire was started with peat or wood, with the draught provided from a water-powered bellows, reaching a temperature that produced purer lead than the bole. Galena and sometimes some coal were thrown into the hearth, and the sulphur in galena was oxidized to sulphur dioxide, a noxious gas. Molten lead trickled onto a metal forehearth and was scooped from a sump into moulds to form pigs of lead. The

85

fire was placed under an arched roof with a chimney and the men worked in an open-fronted building, wearing heavy knitted shawls to protect the head and shoulders from the heat and the strong draughts.

Ore hearths were efficient, but when coal was used its sulphur content lowered the quality of the lead produced, so it could not easily be substituted for scarce fuel-wood. The next step was the development, from about 1690, of a reverberatory furnace in which the coal fire was in a 'firebox' separate from the hearth in which the ore was placed. Flames from the fire passed over the ore under an arched roof to a tall chimney to provide the draught. The heat 'reverberated' from the arched roof onto the ore. Its efficiency was about the same as the ore-hearth, so it was adopted where peat and wood supplies were not so cheaply available. Both furnaces remained in use until about 1900, when new mechanized types were introduced. In the mid-twentieth century, the Imperial Smelting Process was introduced near Bristol to smelt lead and zinc together—a problem which had caused great trouble earlier. It stopped work only a year or two ago.

The serious drawback to both these methods was that the sulphur dioxide, and fumes from lead that had vaporized, poisoned and often killed people and animals, and polluted the land in the surrounding area so that nothing could be safely grown. The partial solution was to build a stone flue to a chimney on a hillside, some of which have survived in the landscape. Some flues were as much as several miles long—that on the moors above Grassington provide a good example of a somewhat shorter one. The lead fumes condensed in these flues and boys were periodically sent in to scrape off the lead dust—a most dangerous task.

The working life of the miner was harsh; some mines had a barracks or hostel, called the 'mine shop', for the miners. In the northern Pennines and Wales especially, it was impractical to walk miles to and from work every day after hours of hard labour, and even more so in winter, so the routine was to walk to the mine on Sunday afternoon after chapel, back home on Saturday morning, and to camp out in the shop for the working week. Each man brought his own food, such as bread, potatoes, tea and sugar, bacon from the family pig, or whatever his wife or mother could provide, though appetites were sometimes poor due to working in bad air. The miners cooked communally and entertained themselves in the evenings by singing, talking, or playing instruments such as the fiddle or concertina. If this all sounds rather charming, the reality was very different.

The restored mine shop at Killhope had four beds for 32 men and boys which, with two shifts, meant three men in a bed and a boy across its foot. Cooking was done in a communal frying pan, swimming with grease, and we can imagine the arguments about whose turn it was to clean it, if it ever was. There was a fire, and the men's wet, muddy clothes were hung up to dry but, as they did so, the fine dust from drilling fell out, covering everything and being breathed in by the inhabitants. (We'll consider lung diseases in a moment.) The window was kept closed against the cold, so it's best not to imagine the odour from cooking, drying clothes, sweat, and other bodily emissions. Washing after work was in a trough or the river, and sanitation was an earth closet, an open latrine or, again, the river. The men may have entertained themselves as described, but it is hard not to believe that fights never occurred. Though Inspectors of Mines came into force in the 1840s for the coal industry, it was not until the later nineteenth century that metal mines were also inspected and conditions were reasonably improved.

The mine also had space, sometimes in the same building as the hostel, for the blacksmith, carpenters, and stonemasons, and for the mine office, where plans, reports, and records were stored. Many of these documents have been preserved and are in the archives of mining museums up and down the Pennines. When explosives came into use from the mid-sixteenth century they were, of course, kept separately.

The miners often had a hard time financially. Craftsmen such as the smith were paid by the day, but for the underground workers the system was that teams of several men would make a bargain with the mine owner or his agent for the next six or seven weeks to drive tunnels at so much per fathom (6 feet), a unit of measure that was widely used, or to produce ore at a given price per bing (a traditional unit of weight) or ton. In Swaledale these rates were called fathom-tale for tunnellers, bing-tale for miners, and day-tale for craftsmen. If the tunnellers struck easy ground, or the miners found rich ore, they could earn well, but the price would be adjusted at the end of the term. The men sometimes cheated by hiding ore for the next period, but the aim of the owners was to enable the men to average, over a few months, a living wage, as it was in their interest to keep a steady workforce. The target in the 1860s, for example, was that a worker was to earn about 15 shillings (75p) per week, but the miners had to pay for candles, tool-sharpening, and explosives, and were only paid for ore that was clean enough after washing to be sent to the smelter. Even the cost of haulage to the surface and bouse-washing was often deducted from their pay. To help them get by between pay-days they were lent money, perhaps £2 per month, which was later deducted. The result was that they could work for months and still finish up in debt—which was, of course, deducted at the next pay-day. On top of that they were unable to work when drought or freezing weather stopped surface working. Sadly, it was actually possible for a man to work all his life without getting out of debt, though matters improved during the later nineteenth century and into the twentieth.

Ill health was tragically a feature of the miner's life. The poor ventilation, damp atmosphere, fumes, and fine dust caused asthma and bronchitis, and it was common for a miner to be 'broken-winded' by the age of 40 and to suffer from the 'black spit', which was probably a form of silicosis, a chronic lung disease for which there is no cure. In the mid-nineteenth century the life expectancy of a lead miner was about 48 years. Another affliction was nystagmus, which is an involuntary movement of the eyes, brought on by prolonged working in poor light. In modern mines, as in the last phases of lead mining, workers are much better protected, with waterproof clothing, steel-toed boots, face masks, hearing protectors, electric headlamps, and hard helmets. The early lead miners, on the other hand, went to work in their own clothes, and even felt caps impregnated with resin were still rare long after their introduction in the mid-nineteenth century.

Away from work, life was not always so bad. Many locally established miners also had enough land for a cow, some pigs, and a vegetable plot. They usually took a week or two away from the mine in September to help get in the hay crop, and to slaughter and salt pigs for the winter. Each family in high moorland areas also had a peat plot from which they got peat for winter fuel, and for the year's cooking, and this fresh-air activity may have helped to improve the miners' health and counteract the effects of mine-work, at least to some extent. The mining/farming pattern shows up very clearly around Grassington and Reeth, for example, where the valley bottom is honeycombed with small fields of pasture and there is a barn to every two or three fields. That seems excessive until one looks up to the fell tops, where the evidence of mining is clearly seen in waste tips. The system was that the man's wife looked after farm and family while he, and probably their older children, worked at the mines. The danger was that a poor farrowing of the pig or a bad hay crop would verge on the disastrous.

The communities in the small villages were close-knit, many of the inhabitants having been there for generations. Entertainment was limited, but brass bands were very popular and were often subsidized by the larger companies. The bands achieved a high musical standard and each had its own semi-military uniform, as elaborate as could be afforded. Methodism was strong from the mid-eighteenth century; the mining museum in Reeth, north Yorkshire, is housed in the former Methodist school room. Naturally, there were occasional festivals and parties, one being the Whitsuntide parade, with bands, chapels parading their banners, and Sunday school queens, the whole affair ending with a tea party and children's games. Communities, and some of the more enlightened employers, often organized their own friendly societies and sick clubs, as a form of insurance, and cow clubs as a means of saving.

The London Lead Company, especially at their Alston mines, was somewhat unusual in being a relatively generous employer, providing cottages with gardens, libraries and reading rooms, and employing doctors and midwives. Schools were supported, with the teacher being paid £100 a year in 1818 when the recommended rate was £24, and selected staff were sent for technical training at the new universities and colleges. A miner was only employed if his children went to school, chapel attendance was almost obligatory, and a miner could be sacked for drunkenness.

In Scotland a report in 1845 into the working of the 1842 Mines and Collieries Act remarked that

> The intellectual and moral condition of this mining population both at Leadhills and Wanlochead is ... above the average. Both

are distinguished by their circulating libraries. The parents appreciate the opportunities of education for their children. There is a
regular minister at each village and the whole population is very
regular in attendance.

which sounds a little too good to be true.

Lead mining declined from the latter part of the nineteenth century
owing to competition from more abundant and easily worked sources
of overseas lead, declining reserves, and high costs. There was considerable emigration, and many mining villages gradually lost much of
their population. Things were not helped by some serious strikes against
the otherwise relatively beneficent London Lead Company and other
lead companies. In one dispute, coal miners from Durham were brought
in, but that proved disastrous as they had no experience of vein mining.
There was some small-scale work in the 1920s and 1930s, but that was
effectively the end of 2,000 years of mining in which uncountable tons
of lead and millions of ounces of silver were produced.

Zinc had been known since very ancient times in India, and had been
traded into the West, but in Britain it was usually used only in the
form of calamine (zinc carbonate), in the manufacture of brass, and
could not be smelted to the metal. There was a calamine mine near
Malham in North Yorkshire. It was not until the mid eighteenth
century that methods of smelting sphalerite and calamine were
evolved by Champion in Bristol, as the uses of the metal become
known. Up to this time zinc minerals had often been dumped on the
tips. The breakthrough in demand for zinc came in the 1820s when
Henry Palmer invented corrugated galvanized iron, though steel is
now used. This new material was cheap, easy to transport, and could
be adapted to local needs, so it was used to build factories and even
Non-conformist chapels, colloquially known as Tin Tabernacles.

The Lead and Zinc Mines

This excellent material was important in Australia and America for roofing houses and barns. It is still widely used for roofs in rural Australia, but also in towns, where it is usually painted. In wartime, huge quantities were needed by the military for camps, as in the ubiquitous Nissen Hut, some of which still exist on old bases.

In more recent times, zinc and its compounds have proved to have a multiplicity of uses as well as galvanizing. For instance, it prevents corrosion of metals exposed to sea water, is used in paints, and even makes organ pipes. Calamine lotion, which contains zinc, is a common remedy for skin rashes. Zinc is also an essential element for life, and zinc deficiencies are associated with liver disease, diabetes, and other illnesses.

A significant source of zinc was the Laxey mine on the Isle of Man, which at its peak employed more than 600 men. The mine also produced lead, silver, and copper. Laxey is also notable for having the largest working waterwheel in the British Isles at 72 feet in diameter. In the Pennines, with the decline of lead production, the London Lead Company was taken over by the Vielle Montagne Zinc Company of Belgium, who worked old spoil heaps for the zinc minerals which had previously been discarded as gangue. They made much use of the flotation process developed in 1860 at the Glasdir copper mine in Wales, where it was found that finely crushed sulphide minerals such as sphalerite are attracted by oils and float to the surface of a tank of water—called a flotation tank. Air bubbles make the process more efficient, a process known as froth flotation. Flotation started to come into full-scale use in the late 1890s, since when huge advances in flotation chemistry have been made. Perhaps ironically, flotation was the death of the British zinc industry: Herbert Hoover, a very eminent mining engineer and subsequently president of the United States, solved the zinc shortage of the First World War by using the method to process old spoil heaps in Australia. After the war, British mines could not compete with such cheap zinc.

Fluorite, commonly called fluorspar, occurred in large quantities in Derbyshire. Small amounts were used in lead smelting, but it came into common use with the growth of industry in the late nineteenth century, especially in America, to which it was exported to be used as a flux to increase fluidity and remove impurities when making high-quality steel. It is also valuable for making opalescent glass, in the refining of lead and antimony, and as part of the production of high-octane fuels. Very pure forms are used in special-purpose optical lenses. Most of these needs in Britain are now met from imports, and the Glebe Mines company at Stoney Middleton in Derbyshire is the last survivor of a once-flourishing industry. Glebe's fluorspar is now used in Runcorn to make hydrofluoric acid as a precursor in the manufacture of asthma inhalers, unleaded petrol, computers, refrigeration and air-conditioning systems, mobile phones, and toothpaste.

A very beautiful blue variety of fluorspar, known as Blue John, makes spectacular ornaments. With the growth of gracious living in country houses and mansions in the late eighteenth century, a small local industry evolved in the Peak District near Castleton and Matlock. The material could be turned on a lathe into vases, made into jewellery, or fashioned as chimneypiece ornaments. A small amount is still produced for jewellery-making, but it is expensive, and antique pieces are prohibitively so.

Baryte is very dense, and when finely ground is used in oil-well and gas-well drilling muds; in the preparation of barium compounds; as a body, or filler, for paper, and cloth; as a white pigment and as an inert body in coloured paints. Barium compounds are used in medicine as 'barium meals' in conjunction with X-rays.

All told, this is a remarkable set of minerals. We'll look at the possible future of lead, zinc, and silver mining in Chapter 14.

6

King Coal: Britain's Powerhouse

Coal was far and away Britain's major mining industry, employing more people and producing far greater tonnage than all the rest put together. Its story features vast output (reaching 200 million tons a year in the late nineteenth century and 287 million tons in the early twentieth), enormous numbers of mines (1,000 were taken into public ownership on 1 January 1947, but many thousands existed over coal's lifetime), the employment at its peak of more than a million men, great feats of mining engineering, and considerable advances in mining machinery. Coal raised steam for engines in mines, factories, locomotives, and ships; it heated homes; made large-scale iron- and steel-making possible; fuelled power stations; provided gas that lit streets and homes, and heated ovens; and it was the source for many chemicals. Some of those uses have had their day in Britain, but coal was the mainstay of Britain's development as a major industrial power. The great railway engineer George Stephenson (1781–1848) suggested that since coal had supplanted wool as the foundation of our wealth, the woolsack on which the

Lord Chancellor sat to preside over the House of Lords ought to be replaced by a coal sack. The proposal was not adopted.

Nowadays, the British coal industry is but a shadow of its former self, with only a handful of underground mines and some open-cast, quarry-like, operations. On the face of it coal's fate is much like that of many other mining industries, which matured, prospered, and then declined, due to factors such as exhaustion of reserves, high costs, competition from overseas, or availability of substitutes. While all these factors, except the exhaustion of reserves, affected coal, the industry also had very difficult working conditions, even worse than those of the metal mines, together with very poor industrial relations and many strikes combined with political influences. Further, it had the problem of methane, a hazardous gas capable of initiating explosions, and the industry could not have grown as it did without a solution to that. But before we unravel those strands it is necessary to say a little about coal and its geology.

One fact about Britain's coal is that we have a great deal of it, from Scotland to Kent; Figure 11 is a map of Britain's main coalfields. Another fact is that there is no such thing as 'standard' coal. There are eight main varieties, ranging from lignite, which contains no more than 65 per cent carbon and so has the lowest heat value, to anthracite, which is 93 per cent carbon, burns without smoke, and has high heating power, but cannot be turned into coke for steel-making. Between these extremes, there are grades of bituminous coal, averaging about 85 per cent carbon, which was once burned in household grates, and some varieties that are relatively easy to make into coke.

Geologically, British coal comes from the Carboniferous system that lasted from 360 to 299 million years ago, the rocks of which are divided into three sub-systems. The oldest is limestone, the shells of

FIG 11 The main coalfields of Britain (approximate)

marine animals laid down over millions of years in warm seas. Next come the sandstones, such as millstone grit, deposited from the rivers of an ancient continent. Finally, there are the true 'coal measures', consisting of seams of coal interspersed with layers of silty rock and bands of ironstone. The coal seams are the result of swamp plant

96

life that decayed and became compressed as coal, but with the seams separated by episodes when rivers deposited silt in the swamp. The seams vary considerably in thickness, and those less than about 2 feet thick are usually ignored, because working conditions are so difficult, unless the coal is of a very high quality, such as the Cumbrian semi-anthracites. However, seams can be as much as 30 feet thick. Seams can be more or less flat, slope at up to 30°, or be steep, as in the north Staffordshire and Midlothian coalfields. Faults are fairly frequent in coal seams and can disrupt a carefully worked out mining plan.

The Carboniferous strata in Britain are generally gently folded, leaving the coal seams in troughs, or coal basins, such as in South Wales, where the collieries on the northern side are mainly anthracite, with bituminous coal on the southern edge. Where millions of years of erosion have scraped away the overlying strata, the edges of the basins are exposed. Elsewhere the coal is concealed by more recent geological systems. About 50 per cent of the great coalfield that runs from Leeds to Nottingham is concealed. The exposed coal shows up in surface outcrops, and on beaches as sea coal that has been washed away from undersea outcrops. The coal can be scraped up from the shore, dug directly from the outcrops, mined from small pits, or via adits that follow the outcrop underground. That was, in fact, how coal mining started. Adit mining was used in the small mines of the Forest of Dean, and in some parts of Yorkshire and Lancashire.

However, nothing in geology is as neat and tidy as this outline of the Carboniferous system. For instance, an old limestone quarry near Ingleton in North Yorkshire has a coal seam embedded in the limestone, and the valley bottom to the west had a small coalfield in the 1880s, while the limestone to the east rears hundreds of feet above the valley. From the Carboniferous series we would expect the Coal Measures to be thousands of feet *above* the limestone. The coal mines on lower ground are an indication of the Craven fault, in which the strata to the west have slid a great distance below the limestone (what geologists call the 'throw' of a fault), and the grits which should be above the limestone have been eroded by ice and water. Over geological timescales, erosion is an incredibly powerful mechanism.

The coal industry is quite old and may date from pre-Roman times, though the Romans mined coal quite extensively throughout Britain for blacksmith work, salt boiling, and heating. The barracks and baths of a fort for 1,000 men required a great deal of fuel, and local sources of wood would soon be exhausted; at Housesteads Fort on Hadrian's Wall a considerable amount of coal was found by archaeologists in a converted guardroom. Villas and public buildings added to the demand, so coal was mined in many areas by bell pits, adits, or open cast, or collected from the beach. In the post-Roman period, a written record of 1214 refers to sea coal being collected in the Firth of Forth and transported by sea to London for heating, lime-making, and metalwork, and by 1300 coal was being dug in England, Wales, and Scotland, for local purposes and for the London market. Attempts were made to stop its use in London because of the smoke, but by then it had become too valuable for brewers, dyers, and other trades. Coal was also used by builders of castles to make lime for mortar. The fourteenth century saw considerable activity on Tyneside, from where coal was exported to Germany.

The coalfields of the north-east had the great advantage of access to the sea and coastal shipping to London, and by 1700 were producing in excess of 600,000 tons per year and employing hundreds of small ships. The insatiable demand for coal brought ever-greater production from deeper mines, and by about 1760 some collieries were 600 feet underground; in 1834 Monkwearmouth in Durham was down to 1,700 feet. Expansion of the coal industry throughout

Britain continued, and by 1800 output was 10 million tons a year, doubling within the next 10 years. From 1870 onwards shafts with depths of over 2,000 feet became common, and Wolstanton Colliery in Staffordshire had, at 3,750 feet, the deepest shaft in western Europe.

From the late eighteenth century canals became important as transport arteries. The Leeds–Liverpool canal increased the accessibility of the growing urban markets, leading to a rapid growth of mining around Wigan and a large rise in the district's population. After about the 1850s the railways, themselves large users of coal not only for fuelling locomotives but also for making rails, expanded at a great rate, which made it possible to exploit the rich coalfields of Leicestershire and Nottinghamshire that had hitherto relied on horse-and-cart transport to reach local markets. The railways at first depressed the coastal shipping trade, but that revived because it was simply more efficient for bulk users such as industry, steam-powered ships which required large amounts of coal for their own bunkers, coaling stations such as Aden which supplied overseas shipping and the Royal Navy, and the generation of gas for lighting towns and homes. On top of all this, British coal was sold to overseas customers. By 1900 this amounted to about two-thirds of the world's trade in coal.

All mining is dangerous, and colliers had to cope with roof falls and the possibility of floods from old workings, and even the sea, which caused many accidents and fatalities, as we shall see in Chapter 11. One extra hazard that was very rare in other mines but common in collieries was methane, which was a problem in coal pits from the very start, causing deaths from explosions. The gas is lighter than air, and burns readily, but if the methane content of air is between 5 and 14 per cent it burns so violently that the shock wave from the sudden flame can blow coal dust into the air and produce a devastating explosion, capable of claiming hundreds of lives in the worst cases. Unhappily, this was not discovered until the nineteenth century, and for hundreds of years men and women underground were faced with a hazard about which they knew little other than that methane, or 'firedamp' as it was called, was dangerous. There were numerous explosions: in 1705 at a pit near Gateshead, five men and a woman were blown clean out of the shaft, their shattered bodies landing some distance away. Sometimes candles were too risky and the men worked by the 'light' from the luminescence of a slab of rotting fish. Where the danger was less great this dreaded gas was burned off at the start of the shift. One man, braver than the rest, would wrap himself in wet sacking, lie flat holding a long pole with a burning candle on its end, and raise the flame into the roof space to burn the deadly gas. He was called the fireman and to this day the man in charge of a coal face, properly called a deputy, is always known as the fireman.

The fireman method would not do; what was needed was a light source that would not cause a methane explosion. Between 1811 and 1820 three people attempted to provide that. One was Dr William Clanny, who in 1813 developed a lamp that used water barriers to separate the lamp flame from the methane-laden air. But, despite later refinements, it was clumsy and impractical. The other two were George Stephenson, the railway engineer, and Sir Humphrey Davy, the most eminent scientist of the day. Stephenson lacked any scientific training but, by trial and error, he devised a lamp in which air entered the lamp via tiny holes and the lamp flame burned within a glass cylinder. He demonstrated the lamp at Killingworth Colliery by holding it in front of a jet of methane streaming from the coal, and proved that the tiny holes prevented the flame in the lamp, even when it had ignited methane inside the cylinder, from passing out and igniting the methane jet. At about the same time Davy was studying the problem. He was the first to prove that methane originated within the coal itself and was not, as had been thought, due to some other source that somehow fed methane into the coal. The flame in Davy's lamp was shielded by a cylinder of metal gauze which, as with Stephenson's, would not allow burning methane to pass out to the atmosphere.

Davy demonstrated his lamp at the Royal Society a month after Stephenson had proved his at Killingworth and a furious row broke out; Davy apparently could not believe that the uneducated Stephenson could have developed the lamp, and he was accused of stealing Davy's design. Both men received financial rewards, but it was not until 1833 that a House of Commons committee ruled that Stephenson and Davy had equal claim to the credit for the safety-lamp design. Davy seems never to have accepted that judgment, and the practical-minded Stephenson thereafter came to distrust theoretical experts. In the event, the Stephenson lamp was used exclusively in the north-east, where the Stephenson family was very influential, whereas the Davy lamp was used everywhere else.

Davy's lamp had two drawbacks: little light could pass through the gauze, and the gauze itself was easily damaged in the rough conditions of a mine. However, as with many inventions, once someone has thought of the basic idea many others refine it, and a rash of improvements soon followed. There were many variations and manufacturers, but the common theme was that the gauze was protected by a metal cover and surmounted a very thick glass tube which, in turn, was on top of the oil tank. These lamps overcame the limitations of the original and not only gave a safe (albeit poor) source of light, but were also an excellent way of testing for methane. If the flame is turned down, a blue cone forms in the lamp, the size of which is an accurate indication of gas content; above 1.25 per cent electrical machinery has to be stopped, while at 2.5 per cent all men have to be withdrawn. Modern mines have air-quality meters and electronic methane detectors.

Another dangerous gas is what miners call 'after-damp' or 'choke-damp': the mixture of carbon dioxide, carbon monoxide, and nitrogen after a methane-induced explosion. Hydrogen sulphide, a highly toxic gas, may also be present. Trapped miners have been slowly asphyxiated by after-damp. 'Damp' derives from fourteenth-century Old German, meaning steam or gas, though the word 'dampf' is used in Danish, which perhaps indicates the very long-lasting influence of the Danish colonization of northern England.

Before we can go further we need to look at working conditions in the early mines, and confront the shame of women and especially children being employed underground. That had been the case since coal mining started, but these were little understood publicly until, in 1842, a Royal Commission was appointed to investigate the matter. The report of the commissioners painted an appalling picture, which was seen at the time as a disgrace to a civilized, and still strongly Christian, nation. Children started work at the age of 8, though as young as 4 was not unknown. The youngest worked as 'trappers' whose job was to pull a rope to open and close a heavy curtain, or brattice, whenever a tub of coal passed. This was lonely work and often done in the dark, and was vital to ensure ventilation, but the 1842 Report said that were it not for the occasional passing of a tub it would be solitary confinement of the worst sort. Older children worked with the women at carting, or 'putting', coal from the face to the shaft. This involved hand-dragging a basket laden with coal in a space only a few feet high, for as much as 200 yards, repeatedly throughout the shift. A band around the head added a little traction but left the head swollen and agonizing to the touch at night, though the task had to be endured day after day. Another method was to harness people, like animals, to a small wagon or tub.

In the coalfields of east Scotland women, who made up about 25 per cent of the workforce, carried coal on their backs up ladders from the coal face to the pit bottom, work which they did quite late into pregnancy. A large basket would be loaded with coal and strapped to the woman's head. A few more lumps were added at the back of her neck and then she set off up the ladder. Coal might well fall from her overloaded basket onto other women on the ladder, and accidents of all kinds were common. The report quotes one woman as saying, 'I wish to God that the first woman who tried to bear coals had broke her back, and none would have tried it again.'

The report formed the basis of the Mines Bill of 1842 outlawing the employment of women and juveniles under 10 years of age underground, though several years passed before the Act took full effect. The 1842 report convinced MPs who passed the Bill without a division, but it was bitterly opposed in the House of Lords. Lord Londonderry, whose income from coal royalties was more than £100,000 a year, wanted to continue with child labour—women did not work in his pits in the north-east—and he persuaded his colleagues to amend the bill so that inspectors could report only on the condition of people in mining, but not on the state of the mines. Seymour Tremenheere was appointed the following year. He was not a mining engineer and did not make underground inspections but acted more as a welfare inspector. In 1850 the Act was extended to appoint Inspectors of Mines, and four of them, all mining engineers, were duly appointed. The mine owners were also required to maintain plans of the workings. All of this was opposed by Lord Londonderry.

Male miners were treated every bit as badly as the women. Generally they were paid on piecework, but usually within a complex system of subcontractors. Wages were better than in agricultural work, but a man's earnings tended to decrease with age as he became less capable of heavy work, so early marriage and a large family were a form of pension. However, wages were often paid in arrears and partly in the form of tickets for the company's 'tommy-shop', where prices were higher than in normal shops. This practice was banned by the Truck Act of 1831, and subsequent measures as late as 1871, but despite the legislation a miner who wanted to keep his job spent at least some money at the company store because, though the owners could not compel him to deal at their shop, neither could he force them to employ him. Not surprisingly, resentment at this sort of behaviour was a factor in the occurrence of bitter strikes and the formation of trade unions, as we shall see later.

With men and women working partly clothed in the hot conditions, debauchery underground was alleged to be frequent and violence was common. The hard labour in cramped conditions over-developed some muscles and stunted growth so that people were deformed and crippled at an early age. It is no surprise that some miners were brutal and vicious, yet the fact is that this life produced some fine individuals, as we shall see in Chapter 13. But the legacy of such suffering, passed as oral history from generation to generation, created an abiding hatred of the worst of the coal owners, though some were relatively beneficent.

The ownership of the coal mines underwent several changes. The very earliest mines, little more than bell pits, may have been owned by wealthy local men or by the gangs who worked them; however, by the Middle Ages much of Britain's coal-producing land was in the hands of the monasteries, who operated coal and iron mines and smelters, thereby generating considerable revenues. For instance, the Monkwearmouth Colliery, mentioned above, was so named because the land at the mouth of the River Wear was owned by a religious house. That all changed between 1536 and 1540 when Henry VIII dissolved the monasteries and dispersed the monks. Some of these valuable properties were given to particular favourites, but because Henry needed money most were sold, in some cases laying the foundations for vast fortunes for the descendants of the original buyer. By about 1850 there were several ennobled families with incomes in excess of £100,000 a year.

However, these families were not always interested in the mines other than as a source of income, and, as the scale of operations grew, finance from local capital was not adequate and the land-owners often wanted to reduce their risks. A natural evolution, especially after limited liability came into force in the 1860s, was the development of colliery companies, which paid royalties to the landowner but were otherwise independent. Iron-founding busi-nesses played a significant role in this. As mentioned above, iron ores are found in the geological Coal Measures, and iron- and steel-making are heavily dependent on good-quality coking coal, so (in what we would now call backwards integration) the iron masters often controlled coal companies, or made substantial investments in them, to ensure supply. This pattern of ownership continued in much the same way until, as we shall see, almost all the coal mines were nationalized on 1 January 1947.

Profitability varied considerably, but as a *very* rough guide, during much of the nineteenth century the price of coal at the pithead was around 5 shillings (25p) a ton but it sold in London, often via agents, for about £1, though that overstates the case because transport costs might be considerable. However, there was substantial volatility; for instance, in 1872 the price of coal soared due to a major strike in South Wales that lasted for 13 weeks and coincided with a big expansion of the Middlesbrough-based iron industry. Matters also went the other way when there was general depression of trade, as in the mid 1840s and at several other points in that century.

Demand is one thing, but output to meet it is quite another, so mining technology underwent radical change in the space of a few hundred years. The early miners used pick and wedge tools with winding by a horse whim or jack roll. As the industry grew, mines went deeper, and by 1490 some Tyneside mines were below the water level and horse-powered and waterwheel pumps came into use. Steam engines followed in the eighteenth century and were used for pumping and for winding coal and people.

Bell pits could not meet demand and mining had to go deeper underground to reach the richer coal seams. One method of underground working was called room-and-pillar, in which numerous headings went into the coal, leaving pillars to support the roof. However, that usually extracted not more than half of the available coal. There were several variations of that method, each with its own name, but the effect was much the same: large amounts of coal were left behind to support the roof, and the deeper the mine, the larger the pillars, which could mean that as little as a third of the coal reserves were exploited. In some cases, when an area of coal had been mined, the miners worked back the way they had come, extracting the pillars as they went, but because the pillars were the roof supports, that was terribly dangerous.

The waste of room-and-pillar could not be sustained, and the solution was longwall mining, first introduced in Shropshire in the late seventeenth century and soon becoming standard practice. In longwall coalfaces a slot, often 200 or 300 yards long, and the height of the coal seam, was opened out and gradually advanced to the boundary of the mine. The coal was undercut by hand and either collapsed under its own weight or was blasted down, after which it was 'filled', or shovelled into tubs and taken from the mine. Each miner had a length of coal to cut or fill—his 'stint'—and the face advanced by about 3 or 4 feet a day. The roof at the face was supported on wooden props, which were extracted as the face advanced. The area behind the face was called the goaf or gob and was supported at intervals by 'packs' built from stone raked from the goaf. This had the effect of controlling the descent of the roof, though sudden collapses could also occur.

Longwall work developed considerably in the later nineteenth century. Mechanical coal-cutters, powered by compressed air or electricity, started to come into use in about 1890, and cut about 5 feet 6 inches deep, as opposed to the 3 feet by hand, but it was a long time before they were commonly used. As late as 1913 only about 9 per cent of coal was cut by machine. Conveyor belts were another improvement on the coal face and eventually extended as far as the shaft bottom, though take-up was slow, mainly because of problems with supplying the necessary power. Steel props and then hydraulic props and steel bars replaced wooden props as roof supports.

Further advances were needed to improve coal-face productivity, and a means had to be found to get away from hand-filling (shovelling) coal. The ideal solution would be a machine that could cut the coal and load it onto the conveyor. Various expedients were tried. One, the Meco-Moore cutter-loader, was an American design, introduced into a British colliery in 1934, though by 1947 only 15 were in use. Its main drawback was its complexity, and a tendency to produce large lumps of coal from time to time, which caused problems on the conveyor. The big breakthrough was the invention in 1952 by the National Coal Board's chief mechanical engineer, James Anderton, of the Anderton shearer. A shearer weighs about two tons and is mounted on an armoured conveyor. Its business end is a drum, fitted with numerous picks, the height of the seam and between 16 and 22 inches deep, depending on the height. As it travels down the face, the rotating drum smashes the coal and an attached plough loads it onto the conveyor. On its return run, the drum cannot cut but another plough loads the remainder of the coal. A short heading is made at each end of the face to allow the shearer to move forward. Much more powerful shearers are now in worldwide use and, in the most modern mines, mechanical roof supports cover the entire face, pushing the shearer forward with hydraulic rams and then, as it were, walking forward to catch up.

Access to a longwall coal face is by tunnels at each end of the face, called the main gate and the tail gate (names derived from Old Norse *gat*, an opening or passage), the former being the air inflow and the latter for coal transport. In early longwall work these were as low as possible, and women and children crawled along them dragging coal in baskets. In more modern mines the gates are about 8 feet high, which involves cutting into the stone of the seam's roof and support-ing it with steel arches. However, the gate roof will gradually be compressed by the weight of the overlying strata, so the foot of each half of the arch is tightly bolted to a wooden prop. The notion is that, as the weight builds up due to the advancing coal face, the props will slide up the arch, despite the bolts. That can happen, but it is also common to see props, about 5 inches in diameter and made of very strong Baltic fir, snapped like matchsticks, and steel arches twisted out of shape, which gives some idea of what roof pressure is like. In practice, roof control on the face and in the gates is a good deal more complicated than I have made it seem.

To reach the coal seams, shafts were sunk, eventually to consid-erable depths. For centuries the sinking was done by hand, though gunpowder was used from about the 1750s. The waste rock was carried or hoisted to the surface. Sinking a shaft that is half a mile deep (and some were considerably deeper), while keeping to a precise vertical axis and maintaining the correct diameter, is not a trivial matter, and the work was often contracted out to a specialist firm. A major uncertainty was that the strata to be penetrated were often not known until the shaft reached them, and it could run into ground that was soft or friable, or that contained water under high pressure. Techniques had to be evolved to deal with those issues, one being to inject the rock with cement under very high pressure to stabilize it. Another is to freeze the troublesome strata and then give the shaft a watertight lining. Modern shaft-sinkers use boreholes to probe ahead, but that technology did not exist before the late nineteenth century.

The problem with the early deep mines was that it was the norm for a mine to have only one shaft, divided down the middle by a wooden partition, but an accident to the shaft left no other way for people to get out of the mine. In 1835 a Select Committee of the House of Commons recommended that two shafts be provided, but it was decades before that became usual, and it was not until the Coal Mines Regulation Act of 1872 that it became mandatory. Indeed, the initial idea was that the shafts need only be 10 feet apart, which meant that an explosion in one could easily damage the other, a result that was little improvement on a single shaft, but by the latter part of the nineteenth century proper arrangements were being made in a spate of large new collieries. The early single shafts had usually been oval, as that gave more space for ventilation to go down and coal to come up, but a circular shaft resists ground pressure more effectively, and circular shafts became the rule, the shafts being lined with iron 'tubbing', segments of a circle overlapping like brickwork. Hoisting of coal and lowering of men had been literally rather hit-and-miss as baskets crashed against each other in the shaft, but the new shafts had proper cages for men, with guide rods and safety features, and cages for tubs of coal, later supplanted by automatically loaded skips fed from bunkers at the end of the conveyor system.

Ventilation, to provide air for people to breath and to disperse noxious gases, was a problem from the earliest mining; in bell pits the air supply was limited. In the early deep mines with a single shaft a common practice was to have one or more furnaces burning at the foot of half the shaft, so that the updraft would drag air down the other half; furnaces were still in use as late as 1880 in some pits, even in those that had two shafts. On top of that, underground steam engines drove rope haulage systems to bring tubs to the shaft. (Even writing about all these open fires in gassy mines makes me nervous!) Ventilation fans on the surface were experimented with from as early as 1768 but they did not really come into their own until electricity became generally available; the National Grid was created as late as the 1930s.

The flow of air through all the access tunnels and coal workings was controlled by heavy cloth flaps, called brattices, which could be opened and closed as required. A whole area of a mine could be shut off if it was not being worked, but that allowed gas to build up in the unworked areas. It was James Spedding who saw that air had to flow continuously throughout the mine, which meant that it had to travel considerable distances through rough tunnels, with a big loss of breathable quality and serious waste of power. The eventual solution was to divide each mine into 'districts'—areas separated by panels of coal. This gave better air quality and wasted less ventilation power, and is now standard practice in most coal mines worldwide.

A significant social aspect of developing and implementing all this technology, including those inventions we looked at in Chapter 4, was the evolution of professional education for mining engineers and others concerned with the mining industries, such as surveyors and mechanical engineers. Much was being done by the early nineteenth century in semi-formal meetings at which papers were presented and views debated on topics such as ventilation and the design of shafts and winding engines, or whatever caught the attention of the participants. These meetings were often supported by, and in turn helped to promote, the emerging colleges of technology at places such as Glasgow, Wigan, Leicester, and Cardiff and just about anywhere else in the coalfields. These mining discussion groups helped to draft legislation designed to improve safety, and in due course they evolved into fully fledged academic departments of mining engineering in the great civic universities of the nineteenth century, which met the statutory requirement that colliery officials had to be properly qualified for their level of responsibility.

The outcome of all these advances in technology and practice was the development of large collieries. To take one example, Bickershaw Colliery was near Leigh in Lancashire. In 1874 two shafts were sunk 1,500 feet deep to reach four seams, one being the rich Pemberton 5 foot seam. More coal lay deeper, so in 1881 two more shafts were sunk to 2,100 feet, and later deepened still further to reach the King Coal and Wigan seams. By 1957, 3,000 men were employed, producing about 4 million tons a year. The colliery was extensively modernized with 7 miles of the most advanced conveyors and coal-handling equipment, allowing Bickershaw to handle the coal produced at two neighbouring collieries. The local community was vibrant and Bickershaw had a famous brass band. Reserves were estimated to be sufficient for another 40 years, but even this big, advanced (by the standards of the day) colliery could not compete with foreign sources. The situation was worsened by the industrial relations problems discussed below. Bickershaw closed in 1992. The site was cleared and remained derelict for years, but the local council plans to create a regional park, with a visitor centre, country walks, and community facilities. *Sic transit* King Coal.

Another social aspect is what we now call 'industrial relations'. It is not easy to find a single word to describe the quality of those relations in the centuries of coal's existence, but appalling and disastrous spring to mind. Given the dreadful working conditions and the rapacity of all too many of the owners, this is not surprising, to which must be added resentment about the terrible accidents that took many lives and which, rightly or wrongly, were blamed on the owners. Attempts to strike and form unions were frustrated by the Combination Acts of 1799 and 1800 which prescribed severe penalties for a worker who 'combined' with another to gain an increase in wages or do anything against his master's interests, though the effectiveness of that legislation is doubted by some historians. There had long been strikes in the coalfields, but when the Combination Acts were repealed in 1824 (though some restrictions remained), trade unions could legally be formed. We can only note some key events.

The early unions were local, poorly financed, and often lacked expertise, so many strikes were defeated, leaving the miners no better, and often worse, off. That was not always the case, as the multiplicity of colliery companies did not always enable them to combine against the workers, and the men won at least some gains for the inevitable suffering of a strike. The local unions gradually combined and the first national union was the Miners Association of Great Britain and Ireland, formed in 1842, the year of the Mines Act. In 1844 the Association led a strike for better wages that lasted for five long months. The strikers endured severe privations, and the autobiography of a collier, discussed in Chapter 13, describes what life was like during twentieth-century disputes.

Despite the strikes, the combination of demand and technology made the growth of the coal industry inexorable. For instance, there were 109,000 coal workers in 1830 but a million by 1913: coal employment rose ten-fold but manufacturing only tripled. This growth of output was matched by increasing numbers of mines. In 1905 there were at least 3,200. Many of those were small companies with limited finance, so there was a slow take-up of new technology, such as electrically powered winding engines and coal-cutters.

The Association was broken up in 1848 by the power of the employers, though there were still successful area unions, and some employers were pleased to deal with an organized union. The area and regional unions combined in 1889 and the Miners Federation of Great Britain, later the National Union of Mineworkers (NUM), came into existence and survived the hostility of the owners. The stronger, and better-organized, Union was able to support disputes, and strikes and lockouts again occurred. A major lockout in 1893 badly affected a number of other industries, one of which was Cheshire salt-making (Chapter 8). There was a good deal of violence; an event that lived long in mining folklore was the 'Featherstone massacre' at Acton Hall Colliery near Wakefield. Troops were called out, two people were killed, and twelve were wounded.

Most disputes were local to an area or confined to an employer, but nationwide strikes or lockouts happened in 1921, and again in May 1926 as the General Strike. The Trades Union Congress called out certain sectors of industry, and others, including the miners, downed tools in support, so the strike spread to affect much of Britain's industry. The government took military precautions, but while the strike on the railways, the docks, and other industries ended after only 10 days, the miners were less fortunate. The government and the owners wanted to make the miners go back to an 8-hour day instead of the 7 hours which had been won in 1919 without a major strike. Not surprisingly, this was bitterly resented by the miners, who stayed out, or were locked out, until November. When they were forced back by hunger, they had to accept longer hours and lower pay, while many were victimized and remained unemployed for years. The hardship that these events caused the mining communities, and the memories of ill-treatment, created a legacy of bitterness that lasted for decades and was an influence on strikes that occurred 50 and 60 years later. Mining communities seem to have long memories, and it is quite common to hear miners referring to events that occurred in the time of their grandfathers as though they had happened yesterday.

The twentieth century was a period of turmoil for the industry. There were serious strikes on top of two world wars and the great depression of the 1930s. The outbreak of the First World War led thousands of miners to volunteer for Lord Kitchener's New Armies. This had a serious effect on coal output, and many of the miners were sent back from the army to the collieries. None the less, coal output had to be cut because the miners could not be provided with food sufficient for their arduous work. In 1916, when the Western Front was almost at a stalemate, many miners, together with sewer workers (who were called clay-kickers for their method of working in clay soils), were recalled to form tunnelling companies in the Royal Engineers. Their greatest feat was the battle in 1917 for the commanding heights of Messines Ridge. In all, 19 tunnels were driven under the German lines, the tunnellers occasionally encountering the counter-tunnels of their German opponents, resulting in vicious underground battles. Many tons of explosives were planted in each tunnel, the detonation of which shattered the German defences. Some of the craters are still visible—yet another illustration of mining's legacy in the landscape.

Between the wars, coal was so important that it impinged on government policy. For example, in 1928, the then Chancellor of the Exchequer, Winston Churchill, concerned that the country spent as much as £45 million a year on imported fuel oil, introduced a budget that included favourable terms for distillation of fuels from coal slack—very small coal that was unsaleable and that was often dumped in the underground goaf, which could result in fires from spontaneous combustion. It was estimated that each ton of slack could produce 4 gallons of pitch for road-making, 5 of diesel oil and

2 of petrol, 2.5 gallons of lubricating oil, a few pounds of candles, and 6 gallons of assorted tars, acids, and other chemicals, from which could be made explosives, synthetic rubber, carbolics for antiseptics, and even umbrella handles. (Humble coal is a very versatile substance!) Unfortunately, the chemical processing was very complex and expensive, apart from all the transport costs, so it was never fully successful. Germany, which had ample coal and whose only oil came from Romania, used a similar process during the Second World War, and improved methods are still in use in India, which has much coal but little oil.

In 1939, with the onset of the Second World War, younger miners were conscripted for military service, and coal output again suffered. An appeal for volunteer miners failed, so about 10 per cent of young men conscripted were chosen at random and, after only 6 weeks training, became coal miners. In all, about 48,000 young men were involved; they were called 'Bevin Boys' after Ernest Bevin, the Minister of Labour in the wartime coalition government. The programme lasted from December 1943 until 1948, but although the Bevin Boys performed vital war service they had no right to return to their former jobs, unlike military conscripts. Their contribution to the war effort was not even recognized until 1995, when the Queen made a speech to mark the fiftieth anniversary of the end of the war, but it took another 10 years for them to be awarded the Veterans Medal, by which time it was too late for many of them.

By 1945 the coal industry was in poor shape. The depression and the war years had starved it of investment, and the austerity after the war—life was even harder than it had been during the war years—meant that there was a relentless drive for coal output to feed the industries that could produce exports to pay off Britain's huge war debts. However, the landslide victory of the Labour Party in the election of 1945 was seen by the miners as the fulfilment of a long-held dream. Nationalization of the mines and other industries had been talked of for very many years and had become a formal policy of the Labour Party; now it could happen. The steel industry and the railways were nationalized, and after the owners had been compensated, the mines were taken into public ownership as the National Coal Board (NCB) on 1 January 1947.

For the ordinary miners, national ownership seemed to have little effect. The owner of the pit was now the nation but the same managers were still there, and the personal relationships and good- and ill-will had not changed. On top of that the NCB required a large management superstructure operating from an expensive London headquarters, via regional divisions and then local areas of a few collieries, down to the individual pits. Large houses were taken over as local offices and many new staff were recruited. To fill all these extra posts was no small task, and (from personal experience) there were cases of 'jobs for the Union boys', some of whom rose to the new responsibilities while others did not. It became almost an article of faith in the NUM that the purpose of the NCB was to provide employment for miners, but the rundown state of the industry after the war meant that very few pits were profitable and large government subsidies were needed, as they were elsewhere in Europe's coal industries.

Other changes, in the early 1960s, were that people and industries did not necessarily want the kind of coal that was being produced, that it could be imported more cheaply than it could be mined at home, and that the government's ability or willingness to subsidize a loss-making industry and to put up with poor industrial relations was not limitless. Change had to come, and one of its agents was Lord Robens, who had been an official of the shop-workers union and a Labour MP. He accepted the chairmanship of the NCB when it was offered to him in 1961 by the Conservative prime minister Harold Macmillan, and made a point of seeing collieries at first hand, visiting about 350 of them. Perhaps surprisingly industrial relations were relatively good during the contraction of the industry in the 1950s and 1960s, and Robens got on well with the leadership of the NUM.

However, during Robens' 10-year term of office, the number of pits decreased from about 700 to no more than 300. Output fell from 194 million tons in 1960 to 140 million at the start of the 1970s, but productivity increased, so that the NCB was finally making a small profit. In 1964 Robens disagreed with the Labour prime minister Harold Wilson, who wanted to accelerate the rate of closures, and subsequently with the Conservative government in 1970, and his reputation suffered from his handling of the Aberfan disaster, which we will meet in Chapter 11. When his period of office came to an end he was unpopular with the NUM. Robens went on to a successful business career and died in 1999.

The closures, and harder attitudes within the NUM, laid the groundwork for some serious strikes, such as that in 1972 which was the first nationwide strike since 1926. The NUM rejected a wage offer that would have left the miners substantially worse off than workers in comparable industries, and the strike began in January—a time of year when the demand for coal was at its highest but also the time when hardship to the miners and their families would be most severe. The miners picketed power stations and then steelworks and other coal users not themselves involved in the strike—so-called secondary picketing; and in February the Conservative government of Edward Heath decreed a state of emergency, and a three-day working week was introduced to save electricity. An agreement was reached between the NUM and the government, and the strike was called off at the end of February, leaving the miners at about the top of the pay range for industrial workers, though within a year they had slipped to a much lower position.

In July of 1974 the NUM demanded a 35 per cent wage increase, which could not be accepted, and the resulting dispute lasted for seven weeks. The Conservative government, still led by Edward Heath, resisted the miners' demands and again enforced a three-day week on industry. There were regular power cuts for domestic users at pre-announced times. Heath called a general election, putting the question: 'Who runs Britain, the government or the unions?' He lost, and the incoming Labour government was more sympathetic to the NUM, partly because the unions were large contributors to party finances.

One of the people to achieve national prominence at this time was Arthur Scargill, who had started his working life as a coal miner at the age of 15. He rose rapidly through the ranks of the NUM as he was an assiduous committee man, a fiery orator, and a diligent trade union official, with a reputation for dealing efficiently with matters such as a miner's widow not receiving her allocation of free coal. The free coal was an important perquisite, and Scargill showed that the NUM looked after its own. These attributes won him such respect among miners and in union circles that by 1973 he was president of the Yorkshire Area of the NUM, becoming president of the whole NUM in 1981. He was also far to the left in politics, which did not endear him to Margaret Thatcher, who became Conservative prime minister in 1979. She saw that the inefficiency of British industry had to be remedied and that the power of the trade unions had to be diminished, and to that end legislation was enacted requiring trade unions to ballot their members before calling a strike, and prohibiting secondary picketing.

The third actor in this impending drama was Ian MacGregor, who was appointed to chair the Coal Board in 1983 at the age of 71. He had enjoyed a successful career in the American mining industries and, on returning to Britain, was successively chairman of the car manufacturer British Leyland and of British Steel. He brought British Steel close to profitability, though at the expense of massive job losses and the desolation of steel-making communities. In his new job, he saw the coal industry as being feather-bedded, and openly said that the NCB's role as what he called 'a social security enterprise' had to end. Scargill's riposte (in an interview given to the BBC on the appointment of MacGregor in 1983) was: 'The policies of this government are clear—to destroy the coal industry and the NUM.' Anticipating trouble, coal stocks had been built up, so that there could be no repetition of the three-day weeks of 1972 and 1974.

In 1983 the NCB announced a further programme of pit closures, threatening 20,000 jobs and several communities, and the brewing storm broke. However, unlike all the hundreds, even thousands, of disputes, great and small, which had marred coal's history, this time it was not about wages, working conditions, or victimization; it was about the preservation of jobs and pits. Since this dispute failed to achieve those aims, we shall look at in a little detail.

Local strikes began, and by March 1984 thousands of miners were out. Strikes accelerated when it was announced that some of the closures were to take place in five weeks' time. Unhappily, the strike engendered much violence, such as that at the Orgreave coking plant, when at least 5,000 miners from across the country were involved in running battles with thousands of police from several forces. There are unproven suggestions of informers within the union keeping the authorities abreast of strike plans, and MI5 monitored people who were using the strike for subversive purposes.

There had been ballots on a local or regional basis but not at national level, so a major issue was whether or not a strike ballot had properly taken place as the law required. Two miners took the NUM to court and won, leading to the NUM being fined £200,000. When the fine was not paid the court ordered its assets to be sequestrated, but they had been moved overseas. A serious consequence of the non-payment of the fine was that the miners and their families were no longer able to claim benefits, causing great hardship. Money was raised but poverty and hunger were, as so often before, rife in the mining towns. Scargill made a tactical error in seeking money from the Soviet Union, and the Trade Unions of Afghanistan (then under Soviet control). The media were generally critical of the strike, though to many miners Scargill was a champion of the working class, and they trusted him to the end, often referring to him as 'King Arthur'.

Despite soup kitchens in the mining areas and nationwide collections of food and clothing, the growing hardship forced more and more men back to work, though these 'blacklegs' were hated by the strikers. A taxi driver in South Wales, taking a non-striking miner to work, was killed when concrete blocks were dropped onto the car. The two miners involved were imprisoned for manslaughter. Another working miner was brutally beaten in Castleford, and there were other such incidents. In some cases families were sundered forever when a striking miner quarrelled bitterly, even violently, with a working relative.

A major factor in the weakening of the strike was the unwillingness of miners in the relatively profitable Nottinghamshire coalfield to support industrial action. Their regional union was expelled from the NUM, leading them to form the separate Union of Democratic Mineworkers, the very name of which was an implicit reproof to Scargill. Another union which did not strike was that representing the colliery officials, the National Association of Colliery Overmen, Deputies, and Shotfirers (NACODS). Had they done so, Thatcher and MacGregor might have been forced to compromise with the NUM, but NACODS' members were worried that a period of inactivity would force mines to be closed as unworkable and unsafe, thereby defeating the object of the strike. (If a coal mine stands idle for too long, roof pressure can crush the access roads and entomb face machinery, and methane can accumulate.) In 1985 two mines were closed for those reasons.

The strike finally ended in March 1985, when the miners marched back to work, heads high behind the union banners and brass bands playing. Women's groups distributed carnations at the pit gates, as the flower is said to symbolize heroism. But the closures continued: despite all the hardships, the strike to prevent closures and save jobs had failed to achieve those aims. A further consequence was massive social and economic problems in the mining communities, where alternative employment was hard to find. South Wales and Derbyshire were particularly hard-hit but few coalfields were unaffected.

The NCB, renamed British Coal in 1987, carried on until 1994, when the remaining mines were sold to RJB Mining under the name of British Coal (now UK Coal), but by then there were only 15 deep mines left in production. Yet, despite all these vicissitudes, King Coal is not only not dead, but may even be reviving a little, as we shall see in Chapter 14.

Despite its colossal scale and virtually nationwide extent, coal mining has left remarkably little legacy in the landscape. The waste tips which were once a prominent part of any coal-mining town have generally been contoured and sown with grass to ensure their stability after the Aberfan disaster, or removed for roadstone. The colliery headgears have been dismantled for scrap or to avoid the expense of maintaining them in safe condition. Similarly, winding engines and colliery buildings have gone, and the shafts have been plugged with concrete. The sites of some collieries have been converted into business parks, shopping centres, etc. Thus the most obvious legacy of the industry is the coal-mining towns themselves, though the houses have been modernized, or even demolished and rebuilt. There are, however, a number of museums. The three most prominent are the National Coal Mining Museum for England near Wakefield, and its Welsh and Scottish cousins at, respectively, Blaenafon in South Wales and Newtongrange, a few miles from Edinburgh. In addition, some collieries and steam engines have been preserved, often with voluntary help, as heritage sites. But, apart from working mines, which are not open to the public, it is almost as though the coal industry had never existed.

The industry did, though, leave another very important legacy in the body of legislation passed by parliament to, among other things, ban the exploitation of women and children as underground workers. Modern equal-opportunities law conflicted with the ban, and the prohibition on women working underground was repealed in 1992. Other legislation culminated in the Mines and Quarries Act 1954, which is 'An Act to make fresh provision with respect to the management and control of mines and quarries and for securing the safety, health and welfare of persons.' These laws specify conditions of work such as 'an adequate and sufficient supply of ventilation shall be provided', which is not very specific. The legislation also defines qualifications to be held by various grades of management and by Inspectors of Mines. The Act is strictly enforced: a colliery manager who loses his manager's certificate for any reason also loses his job. While these regulations still govern mines and quarries throughout the UK, their true contribution has been as a model for similar laws for mining throughout much of the world, with suitable adaptations for local conditions, though certain countries have either not followed that model, or the 'adaptations' have been ignored or become mired in corruption.

Britain's other carbon mineral (excluding oil and gas) is graphite, which is pure carbon, whereas the best-quality coal, anthracite, is only 93 per cent carbon. Unlike coal, which is the compressed remains of ancient forests, graphite was formed about 300 million years ago by hot fluids in areas containing granite—the hydrothermal processes that we saw in Chapter 2. Graphite is soft, cold, and greasy to the touch, and very black, which is why it was used in pencils. Graphite has had many other uses: marking sheep; lining the moulds for making cannon balls; and blacking domestic ironware such as grates and cookers. It is also a lubricant, and powdered graphite is ideal for 'oiling' door locks. Other modern uses include 'brushes' in electric motors, and making small crucibles.

Graphite is sometimes found as traces in the Cornish granite, and small amounts were produced in Glen Strathfarr in Scotland, but Borrowdale in the Lake District was the main source of this valuable mineral. Locally it is called wad, though in mineralogy wad is a manganese mineral. Perhaps the name derives from German, as German miners were imported to Cumbria's mines. Wad has probably been used since ancient times, but the earliest records date from 1540. Early works seem to have been small-scale for local use, but the industry took off from 1752 when John Bankes started the Keswick pencil industry, though his family had worked wad mines from about 1635, and continued to do so for some 200 years.

Wad occurs in veins, usually as more or less random nodules, but also in much larger concentrations, known as sops, which may contain several tons of graphite. This haphazard pattern of occurrence was a perpetual bane to the wad miners as they never knew where the next nodule or sop might be, and the workings meander a great deal in trenches or underground. Of course, the random nature of the nodules worked both ways, and in the 1760s seven tons of graphite were extracted in 48 hours from one sop, which was worth the enormous sum of £3,000 in 1760 money. Such potential riches were attractive to the nefarious, as even a pocketful of wad would be very valuable to a hard-pressed miner. There was a high ratio of supervisors to miners, and the latter were often searched as they left work. Armed guards were employed to guard the stocks of wad at the mine, and armed escorts took it to the pencil factory in Keswick. On at least one occasion guards and thieves exchanged shots. One William Hetherington was particularly crafty as, from his own small copper mine, he had a secret tunnel into his neighbour's wad mine. His reward was to be made manager of the wad mine, perhaps on the principle of setting a thief to catch a thief. Theft of wad was, however, generally treated seriously, and could earn the malefactor transportation to Australia, though that was common for many crimes in the late eighteenth and early nineteenth centuries. One particularly notorious thief, known as Black Sal, is said to have been hunted to her death with hounds.

The wars with revolutionary France at the end of the eighteenth century cut off the French from their supply of graphite. In response, in 1795 Nicolas-Jacques Conté invented a process using poor-quality French graphite, powdered and mixed with clay, to make what we now call pencil leads. This enables them to be made cheaply and to any required degree of hardness, and was a major blow to the Lakeland pencil industry. Artistic-quality graphite was still in demand, but wad mining ceased in 1891 when the deposits were exhausted. Traces of the trenches and workings can still be seen, but it is worth repeating that old mines are extremely dangerous places and should not be entered. Pure graphite still has many industrial and artistic uses, and is now imported, principally from Sri Lanka. The Lakeland pencil industry still flourishes using synthetic materials.

7

Building Britain: Granite, Stone, and Slate

People need shelter from the elements (or from enemies), and places in which they can live, work, and rest. While dwellings can be made from wood and straw, stone is far more long-lasting, robust, and fire-proof. Stone has been used since ancient times; the oldest habitable such buildings in Britain being the brochs of Scotland, some of which date from around 600 BC: about 700 seem to have existed. These are beautifully built tower-like, slightly conical, structures which are the finest examples of the craft of dry-stone masonry, the stones being so cunningly laid that mortar is not required. Internal partitions accommodated family groups in rather smoky discomfort. Most brochs are now in ruins or have vanished, but some fine examples survive in the Shetlands and the Outer Hebrides.

One of Britain's most important building materials is granite, an igneous rock created by the cooling and crystallization of magmas extruded from deep below the surface. Some of these extrusions are vast: Dartmoor is one huge mass of granite, now eroded at the surface. True granite consists primarily of quartz, feldspar, and mica. There are several varieties of feldspar and which one is present governs whether the rock is 'true' granite, or one of the twenty or so granitic rocks. All of these are capable of supporting great weights and tolerating weather erosion and hard wear, and the finest quality can be used for statuary. The stones display a wide range of colours ranging from a near brilliant white to dark red. Adjacent quarries at Shap in Cumbria provide pink and blue granites. Interestingly, planetary study of Mercury, Venus, and Mars suggests that Earth is unique in having granite, probably because these planets do not have plate tectonics.

The use of granite has a long history: for instance, the ancient Egyptians mastered the art of carving this tough material into highly polished statues. Closer to home, granite is widespread in the western portions of Britain, Scotland being famous for the quality of its granite. The finest curling stones are made from Ailsa Craig granite, but Aberdeen was Scotland's principal source, due to a combination of cheap labour, high-quality stone, and easy access to the sea for transport to London and other east-coast ports. However, granite has also been exploited commercially on Anglesey, in north and east Wales, Cumbria, Cornwall, and Devon.

Granite has many uses. The poorest quality is crushed and used in road-making. The next best is carved into more or less cubical blocks, about six inches in size and called setts, for pavements. Massive blocks of better-quality stone have great structural strength; the Thames embankment and London bridge are built of Aberdeen granite, as are Sheerness and Aberdeen docks, and many other major structures. Finally, a combination of the best-quality material and great skill creates beautiful features such as the fountains in Trafalgar

Square, statuary, and monuments. A good deal of polished granite is seen in cemeteries. We'll mainly look at Aberdeenshire, since it was the dominant area for granite quarrying.

In medieval times, Aberdeenshire buildings such as castles were built from granite picked up from the ground, and the first proper quarry did not start until 1602. Growth really came in the 1740s, when a serious fire in Aberdeen, which was mainly built of wood, led the City Fathers to decree that granite was to be used for major buildings. To supply it, the early quarries were simply dug into a hillside. The stone was broken using black powder in holes about one inch wide and five feet deep. That, however, tended to produce small stones, and it was not until 1819 that John Gibb who, with his sons, became prominent in the granite trade, used larger drill holes, 2.5 inches wide and 12 feet deep, to 'heave' the rock in large blocks. The eventual technique was to blast in three stages, once to create cracks in the rock, secondly to loosen it, and finally to bring down large slabs or boulders. From the late 1860s steam-powered boring machines capable of drilling 30-foot holes came into use. However, as we shall see later with slate, much depended on the quarryman's 'feel' for the rock in deciding exactly where to place the charges. Blasts often brought down 100 tons or more, but these only happened at specific times, as it was common to have several gangs working in the quarries at once, and it would have been chaotic and dangerous for each gang to blast whenever it liked.

In the early quarries, the small slabs were hand-carried to the top and taken by horse and cart to Aberdeen. Slabs large enough for docks and bridges needed steam cranes to hoist them, which involved dismantling and moving the crane as work progressed around the quarry. The clever solution from about 1870 was to leave the steam engine in one place and to have a network of cables covering the site, with travelling cradles that could lift and transport blocks of granite weighing up to 5 tons from anywhere in the quarry. These were called 'blondins', after the daredevil who had crossed Niagara Falls on a tightrope, and these technological innovations generated a sophisticated engineering industry in Aberdeen. The drawback to the blondin was that a man had to go out along the cable to free a jammed pulley.

In the early days, when a quarry became too deep to work, or water was too great a problem, it was often easier simply to abandon it, move to another outcrop, and start again, but by the 1870s the larger quarries were run on more efficient principles. That involved careful planning of work, with the overburden of soil and eroded stone being progressively shifted as the quarry grew. Development was greatly assisted when transport to Aberdeen improved with the new turn-pike roads, a canal, and eventually railways, so that from about 1830 onwards new quarries were developed northwards to Peterhead, which has an excellent harbour, and inland, and the Gibb family developed the great Rubislaw quarry. The net effect was that by 1870 output was 100,000 tons per year. There was a surge with a building boom in the 1890s when the quarries employed about 2,000 men. Thereafter there was steady decline as architects called for a greater variety of colours for stone facings on buildings, and some beautiful granite-type stones were imported from Norway. There was a short-lived boom in the 1930s, though by then employment was down to 700, and there was competition from Czechoslovakia. Reserves in the best quarries became exhausted, and nowadays the quarry industry is virtually dead.

The quarries, of which there were many, some employing only a few men, were on land leased from its owner, but unlike almost all mining and quarrying, this was usually not on a royalty-per-ton basis: the landowners preferred a fixed rent because the land was very poor for agriculture. Needless to say, the quarrymen worked in the open in horrible conditions. Accidents occurred as, even with their great skill, it was not possible to predict precisely how the rock would break. Wages were very poor: about 1 shilling (5p) a day for a six-day week in 1750, which had risen only to about £1 per week in the 1820s, 70 years or some two working lifetimes, later.

The granite industry would have been nothing without the stonemasons of Aberdeen. Their role was to make the rough granite that came from the quarry into building stone or paving setts, though these were separate trades, each with its own apprenticeship. The work was done by hand and eye using a hand pick, until pneumatic tools became available in the 1890s. Large boulders from the quarry had to be sawn into regular blocks by hand, until steam and electric saws were developed. A long blade operated by two men was used, with iron- or emery-oxide powder as a grinding agent between the blade and the stone, copious flows of water cooling the blade. To cut a facing slab measuring 8 by 4 feet could take four months of unremitting labour. Steam saws reduced the task to 10 days, though hand sawing continued until the 1890s. It is a tribute to these men's skills that when Shap stone was used to build Southampton docks in the 1930s, the dock walls were laid to a tolerance of a minute fraction of an inch to make them watertight.

For decorative work, granite was turned on a lathe to make pillars for buildings, often with a cylinder cut out to leave a hollow core. That reduced weight and saved money on expensive granite, the cylinder being sold as a roller. The difficulty was that the heat generated from friction on this hard stone was so great that the cutting tool soon lost its 'temper', or toughness, and became softer than the granite it was supposed to cut. The ingenious solution to that was a rotating disk of tool steel that had time to cool before it next came into contact with the stone. Another triumph of the nineteenth-century Aberdeen masons was to produce highly polished slabs for the fascias of buildings and for gravestones, though cremation reduced the latter demand. Their final achievement was polished granite statuary, recreating what the Egyptians had done a few thousand years earlier.

Although Aberdeenshire produced granite for all uses, very large amounts of a granitic rock called porphyry came from Wales. The main area was Penmaenmawr on the north coast near Conwy, where a great headland rises from the sea. The industry started with boulders picked from the beach and used locally, but in the eighteenth century a great quarry was developed to supply roadstone and paving for Lancashire, the English Midlands, Rhyl, and Llandudno. Hundreds of thousands of tons were quarried until the top of Penmaenmawr had been sliced off. The men earned 15 shillings (75p) a week working at altitudes of a thousand feet above sea level, sometimes in the teeth of gales from the Irish Sea. The Aberdeen sett-makers sat down with the slab on a bed of sand and grit in an old barrel, but the Welsh way was to hold the stone under one foot and bend down to work it. Acute curvature of the spine was an inevitable consequence, and older sett-makers were referred to by a Welsh phrase meaning 'both ends meet'.

In about 1820 the Scottish engineer, John Macadam, discovered that cubes of stone about two inches in size made a very effective bed for the new turnpike roads. There was soon a huge demand, and large amounts of 'macadam' were produced by smashing granite with heavy hammers. Mechanical crushers were more efficient but cost jobs. Modern 'tarmac' uses any suitable stone, sealed with asphalt, but still based on Macadam's principle.

Most parts of Britain have good building stone. Flint was used in the chalk areas of southern Britain, and some of the older houses in

Wiltshire were built with chalk blocks which stand erosion surprisingly well—though some varieties of chalk are fairly soft, so the builders had to choose carefully. A few other examples: Cheshire has red sandstone; Oxfordshire uses a honey-coloured stone; the centre of Alnwick in Northumberland is built of local sandstone ashlars, carefully dressed blocks, measuring 18 by 9 inches, which gives very pleasing proportions; and quite a few buildings in northern Northumberland have used stone 'quarried' from Hadrian's wall.

However, one of Britain's most beautiful and widely used building materials is limestone from the late Jurassic or early Cretaceous systems, and is about 170 to 100 million years old. The main quarrying districts were the Isle of Portland; near Shepton Mallet; just outside Bath; at Box, east of Bath; and Purbeck, near Swanage. Each of these places had its own characteristic colours and quality. Bath stone, for instance, is relatively soft and easy to work but does not withstand the elements as well as Portland stone, which is harder and more difficult for the mason to dress and carve. Purbeck stone is a shelly limestone which can take a very high polish, making it look like marble. The best qualities of limestone are called freestones, meaning that they can be cut in any direction without splitting, making them ideal for carvings.

Many notable and beautiful buildings came from these quarries and mines. Salisbury Cathedral is built of 70,000 tons of Purbeck stone. In London, St Paul's cathedral, the Banqueting House and much more in Whitehall, Broadcasting House on the Strand, Wren churches, and very many others buildings are of Portland stone. Portland also provided the material for some magnificent country mansions, as well as some 800,000 headstones in Commonwealth war graves in France. Wells cathedral was built in stone quarried at Doulting near Shepton Mallet, and the Regency city of Bath is of local stone. Most of Regency Bath was built speculatively, so while the fronts of the terraces are beautiful, the backs are of rough stone.

Throughout these limestone deposits in Dorset and Somerset there were many quarries and underground mines, some small but others quite large. They declined or prospered as reserves were exhausted and new ones found, or as the demand for the type of stone they produced waxed and waned with architectural taste. Cost frequently meant that a building might be faced with Portland stone, rather than built with blocks, so the masons had to be capable of making heavy blocks or facing slabs to tight tolerances, as well as carving ornamental additions. The masons worked entirely by hand for many centuries, their apprenticeships and conditions having been defined in detail by Charles II. As time passed, large masonry works were developed, with increasing mechanization for sawing, shaping, and turning. Fortunately, the art of the mason has not been lost, and York Minster and Salisbury Cathedral, among others, support schools of masonry to meet the continuous task of maintenance and restoration of weathered stone and carvings.

The numerous and dispersed limestone workings were not efficient, and, starting in the early nineteenth century, very prudent amalgamations and takeovers led to the formation of the Bath and Portland Stone Companies Ltd, a prosperous and well-managed firm. Nowadays, the decline of the industry has left two separate businesses in Bath and on Portland.

Work in the quarries was, of course, hard, dangerous, and dirty, and was done in all weathers, but much stone was mined underground at Bath and at Box in Wiltshire, where high-quality stone had been discovered by Isambard Kingdom Brunel (1806–59) when driving a tunnel for the Great Western Railway. Underground mining had the advantage of producing stone that had not been weathered, and it was not necessary to remove the overburden of soil and loose material, but it was hard labour. An access tunnel about 25 feet wide was driven from the shaft. At the face a slot about 5 feet deep was cut across the top by hand pick, but in order not to waste good stone it was only 9 inches deep at the face, narrowing to 4 inches at the back. This meant that the miner had to swing the pick so as to throw the point into the slot, and it was the most hated job in the mine. Once the slot had been cut the stone was sawn, by hand, from the slot to the floor to define three or four blocks. The left-hand block was then cut out by driving wedges across the base to split it from the floor and the back of the block. Holes were drilled into its face for attaching hooks, and the block, weighing between 5 and 8 tons, was dragged out with a ratchet and chain. The miners now had access to the sides of the remaining blocks, which were carefully sawn so as to avoid cracks or shelly beds. Fortunately, the roof was strong and did not need support. The work was done entirely by hand until 1948, when an adapted coal-cutter was installed. Eventually, 100 miles of galleries were driven between Box and Corsham, and these were used during the world wars for storing ammunition. The Bath stone mines were worked similarly, though they are now causing concern about subsidence as the city has expanded. The Museum of Bath at Work has a recreation of the miner's labours.

Overall, Britain has been as fortunate in its building stone as it has been in metals and coal. The true legacy of the stone industries is the palatial houses and cathedrals which beautify the land and enrich our lives.

Limestone has very many more uses: in steel-making; for cement, roadstone, and railway ballast; and for neutralizing acidic soils. In the nineteenth century the Halkyn lead/zinc mine in Flintshire produced up to 100,000 tons per year of high-grade limestone from under-ground galleries up to 5 miles from the shaft bottom, and there were stone mines at Dudley in Worcestershire. Vast amounts of limestone have been, and still are, quarried in the Mendips, Derbyshire, and North Yorkshire, though the quarries are often in National Parks. We will look at this issue in the final chapter.

Slate was an important industry, and it is a truism that, if the Industrial Revolution was driven by cotton and steam, it was roofed by slate. Slate is mudstone—soft, fine-grained, sandstone—that has been affected by high temperature and high pressure. The colour is black, blue, purple, red, green, or grey, and it can take a high decorative polish. It can even be enamelled to give yet more colours. Slate is found in many parts of Britain, such as the Lake District and Devon, but the main slate area was Gwynedd in north-west Wales, which produced some 90 per cent of British output. One problem in slate mines and quarries is that geological forces have distorted the mudstone layers so that they can stand on end, slope at inconvenient angles, or be twisted. Fortunately, slate has planes along which it is easily split to make roofing slates, headstones, bookends, house name plates, paving slabs, formerly writing slates for schools, and so on. Furthermore, while mudstone itself is virtually useless, slate is highly durable. When restoration work was done in the 1930s the slates laid on the roof of St Asaph cathedral in the late sixteenth century were found to be still useable after about 350 years of Welsh weather.

Slate was used by the Romans and in medieval times for castle roofs and water cisterns, but it took until 1765 for it to be exploited on a large scale by Richard Pennant in the Penrhyn quarry at Bethesda, a few miles inland from Bangor in North Wales. Pennant had made a fortune from slave plantations in Jamaica, and used it to finance the ownership of the land and the development of the slate. In 1873 he was made Baron Penrhyn in the Irish Peerage, taking the name from Penryhn in County Louth, but the peerage became extinct when he died in 1808. It was recreated in 1866, and Edward Gordon Douglas-Pennant became the First Baron Penrhyn of Llandegai, and it is he who features later.

Once Penrhyn quarry had started, others soon followed, such as Dinorwig near Llanberis, which was eventually to be greater than Penrhyn in terms of numbers employed and physical size. In all, some 200 or 300 quarries, many small, came and went over the next two centuries. Some were true quarries, while others were underground mines, such as Llechwedd and Oakeley near Blaenau Ffestiniog, which was to become a major centre for slate.

The quarries were worked in a series of galleries or terraces going up to the surface. A gallery was typically 50 or 60 feet high and anything from 6 to 60 feet wide, growing wider as work progressed. The complexity of producing slates needed several categories of worker. First were the true quarrymen, who saw themselves as the princes of the craft, and worked in teams. Two members of the team dangled from the top of the step on ropes or chains and drilled shot holes to loosen the slabs. This required great skill and an 'eye' for slate to determine how to get the best-quality material—they said that slate only understands Welsh. Every hour a bell was rung one minute before blasting time, after which another bell rang four minutes later as the signal to get back to work. The third man in the quarrying team worked on the gallery, dressing the blocks for transport to the team's slate-splitter. The second group of workers were 'bad rock' men whose job was to extract slate that was not valuable. Another group, the rubbish men, moved waste from the quarry to the waste heaps. Finally there were the general labourers and the craftsmen. The men used a 'caban' (a sort of cabin) for meal breaks and for heat and shelter in the very worst weather, and it is reported that these times were marked by political discussion, union meetings, and the recitation of Welsh poetry.

The men were employed on monthly contracts, though the 'quarry month' was four weeks, and worked a given area at so much per ton. Theoretically that price was subject to negotiation, but the men often had little choice but to accept what the quarry management offered. They received a weekly 'sub' with a bonus at the end of the month for what they had produced. However, after the men had paid for explosives, and even the air for pneumatic drills, the 'bonus' could turn out to be zero. Working conditions were very poor and many of the men suffered from occupational diseases such as silicosis from dust, hernia from levering heavy slabs of slate, and stomach problems from the poor diet—many men drank only very strong, stewed tea. The slate splitters often had haemorrhoids from sitting for hours each day. Men from distant villages, or those who travelled from Anglesey, lived in barracks during the week. Living conditions, especially in outlying villages, were worse than in Victorian slums, and life expectancy was, as was common in the mines, about 47. Other than domestic service at a great house there was no paid employment for women.

However, the main interest of the history of slate is in its cultural, religious, social, and political aspects, which were powerfully inter-twined. Gwynedd was, and to a large extent still is, mainly Welsh speaking. Most of the modern inhabitants are bilingual, but even into the twentieth century 70 per cent of the slate workers spoke only Welsh, and others disliked speaking English. In religion, which was a powerful influence at the time, they were strongly non-conformist, favouring the Congregational and Calvinistic Methodist chapels. This influence is clearly seen in place names, in which the original Welsh was changed to biblical references such as Bethesda, Bethel, and many others. The miners and quarrymen were also fiercely, and rightly, proud of their craft skills; slate-splitting contests were part of local eisteddfods (festivals), and it was regarded as a right for a son to follow his father into the quarry. They also jealously guarded what they saw as the independence they gained by working in teams on their own stretches of slate.

Politically, the local people tended to be Liberal and Radical, inveighed against the established status of the Church of England in Wales, and sought civic equality for non-conformists and social equality for tenant farmers and labourers. There was great resentment that Welsh land was owned by Englishmen, leading to the first stirrings of Welsh nationalism. After the Reform Acts of 1867 and 1884 the Liberals gradually gained political power by winning what had been safe Tory parliamentary seats, though they were later accused of doing little for the miners and quarrymen when industrial strife occurred. These social forces had their effect in the formation in 1874 of the North Wales Quarrymen's Union, which was anathema to Lord Penrhyn, who was willing to provide a hospital and some pensions, but who was bitterly opposed to what he saw as radicalism.

The other side of social history was the families who owned the land on which the quarry or mine lay. Some of these were local people operating small quarries employing only a few men, but others, such as the Assheton-Smiths, Lord Penrhyn, and the Greaves family, were English, and some were very rich. In 1859 Lord Penrhyn drew £100,000 a year from the labour of 2,500 men in Penrhyn quarry, the men earning about 12 shillings (60p) a week on which they might have to support a large family. The managers and agents were also non-Welsh, did not speak the language, were Anglicans and Tories; and in the politics of the time a Tory was nothing like a modern Conservative. The Tory ethos was the right of landowning—in the late nineteenth century about three-quarters of Carnarvonshire was owned by five families—and the corresponding power of master over man. It also meant protecting and, if possible, enhancing inherited wealth, which implied strict control of costs and the search for more efficient methods than separate gangs.

The inevitable consequence was a series of bitter strikes and prolonged lockouts at Dinorwig, Llechwedd, Penrhyn, and many of the smaller mines and quarries. The ostensible cause was often an injustice such as victimization of someone who had been too outspoken, or the refusal by management to re-employ strike leaders, or it might be a wage demand, but the underlying tension was contributed to by management attempts to impose new working rules. The management rationale was that the working of a quarry as large as some of these had become has to be planned years, or even decades, ahead, and that was not consistent with many small gangs working in their own fashion all over the place. In effect, they wanted to replace contract working and bargains with a more factory-like, regulated, wages system, which was much resented. We will let the Penrhyn lockouts stand for the rest.

In late 1896 the Penrhyn lodge (branch) of the quarrymen's union sought a wage rise from Lord Penrhyn. Attempts at negotiation failed, and on 28 Sept he suspended the union officials. The men struck and were locked out. Bethesda was virtually a one-quarry town with little alternative employment, so many men left to seek work in coal mines or in Liverpool, which had a large Welsh-speaking community. The dispute ended in August 1897, nearly a year later, in a complete victory for Lord Penrhyn. However, 1896/7 was small beer compared to the lockout of 1900 to 1903. Again, the immediate cause in early November 1900 was something fairly trivial, but there was some violence, the police were called, and the arrested men were committed for trial at Bangor. On the appointed day, the quarrymen marched there in such numbers that 300 cavalry were put on standby. Later that month 2,800 men walked out and were then locked out. Lord Penrhyn wanted to defeat the union because he saw it as a threat to his power and status.

By June 1901 some men had been forced to seek work, which they only got if they were loyal to Lord Penrhyn. These 'blacklegs' were hated. Troops had occasionally to be brought in to escort them to work, as the local police were unable to cope. The blacklegs were hooted at in the streets, shops refused to serve them, and in many cases they were driven from their chapels—which, since religion was deeply embedded in their lives, was hard indeed. The only alternative was Anglicanism, which was Lord Penrhyn's church. Bethesda's many chapels declined, and continued to do so even during the religious revival which swept Wales in 1904–5.

The loss of wages during the lockout caused terrible hardship. Funds were collected at other quarries and industries, some help came from the trade union, the rich Welsh chapels in Liverpool raised money, and the Bethesda choirs brought in £33,000 from a singing tour. It was not enough, given that 2,000 men out of work implying about 15,000 mouths to feed. By the end of the dispute in November 1903, after three full years, famine was a real possibility. A contemporary Welsh hymn sings of seeking God's help 'in the face of all suffering that is or is yet to come'. Moving words indeed.

Inevitably, men with families to feed and clothe, but unable to face the opprobrium of being a blackleg, left the town for work else-where, or even emigrated, and it took many years for Bethesda to recover. When the dispute ended in victory for Lord Penrhyn about a thousand of the original workforce were re-employed, but only those loyal to the Penryhn interest. Many new men were recruited. An ironic consequence was the import of American slate and growth in the use of roof tiles.

The later history of Welsh slate is swiftly told. The First World War just about destroyed the industry. Export markets were lost, skilled men died in France, and the quarries and mines were starved of investment. There was a short-lived post-war building boom, but the slate industry declined to become a shadow of its past glory. Slate is, however, still produced in North Wales, Cumbria, and Devon. The National Parks, which demand that builders use traditional materials, have been something of a lifeline for slate, stone, and granite. The landscape legacy of slate is the quarries themselves, and the great heaps of waste that surround them. As well as the National Slate Museum in Wales, several old quarries are now tourist and educational attractions.

Two slightly less glamorous industries are clay and gravel working. Clay has been used since very ancient times for brick-making, and large pits remain to mark its existence. It is a very versatile material yielding everything from house bricks to refractory brick used in furnaces and to high-strength engineering bricks. Clay also played a major role in the building of the canals during and nineteenth century, when the canal sides were lined, or 'puddled', with clay to prevent leakage; Leicester Blue Daub clay was the top-of-the-range material. Gravel is also an essential component of building, and has been used as a base for roads since Roman times. Its main modern use is in concrete, which was also known to the Romans. Abandoned gravel pits in Hampshire are now used by the Forestry Commission as sustainable woodland. The Cotswold water park is an area of water larger than the Norfolk Broads, but is simply flooded gravel pits.

China clay is not a building material but it has provided much that is beautiful and utilitarian in our dwellings, and has remained a big business in Cornwall and west Devon. It has been a major feature in the shaping of the regions and their landscapes, but, unlike the south-west's copper and tin mining, which date back 3,000 years, the china clay industry started as late as the mid eighteenth century. The driving force was the fact that, while fine china was highly prized by increasingly wealthy people, it was very expensive, and the supply from China, where it had been produced since about AD 800, was uncertain in the sailing conditions of the day. That started a search for china clay, and it was finally discovered in west Cornwall in 1746 by William Cookworthy, an apothecary. It was subsequently found, and mined, on Dartmoor and Bodmin Moor and near Lands End, but principally around St Austell.

China clay is formed by the decomposition of the feldspar in granite, by hydrothermal activity, and erosion by water over millions of years when the climate was warm and wet. It is a white mineral, also called kaolin, the name deriving from a region in China. Kaolin has the useful property that, when mixed with about 25 per cent water, it can be pressed into the shape of, say, a cup, and it retains that shape when fired in a kiln. The cup, or other form, can then be painted and fired again, perhaps several times, to produce high-quality and expensive china and porcelain.

Pure china-clay particles are exceedingly small—a few millionths of a metre—which makes it very valuable in paper-making. About 30 per cent of writing paper is actually kaolin, but kaolin is also used in paint-making and to add strength to car tires. Nowadays, about 80 per cent of china clay is used in paper-making, 12 per cent for pottery, and the remainder in paint, rubber, plastics, cosmetics, and pharmaceuticals.

A china-clay deposit contains on average only about 10 per cent of kaolin, together with 45 per cent of unaltered granite, the remainder being mainly sand. It is mined from so-called pits, which are actually quarries, some being very large indeed. These are a striking feature of the landscape around St Austell and in west Devon. For hundreds of years the work was done by hand, and, the climate of Cornwall being somewhat variable, the working conditions, as with so many of the old mining industries, were bad. The old clay workings left very distinctive white cones of waste, but the modern industry takes enormous pains to restore the landscape when china-clay extraction finishes.

Unlike most of the mining and quarrying we have looked at in this book, the china-clay sector still thrives, though modern pits are fully mechanized. First, the topsoil is removed and set aside for later landscaping. The clay and gravel is then broken by very powerful water jets, called monitors, and washes down to the lowest part of the pit, from where it is raised to the surface by pumps capable of handling stones 6 inches in diameter. Waste rock, called stent, is taken away by truck. Alternatively, the rock is broken by explosives or by powerful diggers and taken up by truck. At the surface plant, a very complex and sophisticated sequence takes place. Large granite boulders can be used for sea defences, but everything else is crushed to small sizes and washed, with water being recycled, and the kaolin is processed in various ways to meet the needs of different industries. Output is about 3 million tons per year, 80 per cent of which is exported. Sand and gravel are sold as by-products to be used for building in Cornwall. If transport costs could be reduced, they would be sold more widely.

Waste is dumped in old pits for eventual landscape restoration, and the very fine sand and mica accumulates in lagoons. When these are full and dry the surface is seeded—one of these is now a football pitch. Old pits are used as reservoirs, but also provide facilities for fishing and for tourists. The famous Eden Project is in an abandoned china-clay pit. If you can tear yourself away from the domes showing the plants of the world, stand and look up to the top of the pit and imagine what it was like to work there a century or more ago. There is a china-clay museum at Wheal Martyn near St Austell.

8

The Salts of the Earth

Without some salt in our diet we die, so production of common salt, sodium chloride or halite, is vital. But halite is only one of a group of chemicals—the salts—many of which also have very valuable uses. One such is potassium chloride or sylvite, and Britain's halite and sylvite are the sources of chemicals that flavour our food and fertilize our fields, but are also used for making glass, soap, paper, explosives, bleach, detergents, margarine, dyestuffs, plastics, and much else, as well as being used on the roads in winter. Yet another salt is anhydrite, calcium sulphate, which can be used for making sulphuric acid, with its many uses, such as making ammonium sulphate fertil-izer. 'Anhydrite' means 'without water', but its relative, gypsum, has two water molecules attached, and that makes gypsum useful to builders as it can be made into all sorts of plaster products. Alum is chemically more complicated as it is potassium aluminium sul-phate, but it was once commercially valuable as it makes dyes cling to fibres. All of these are the evaporites that we mentioned in Chapter 2 and that were built up between 250 and 200 million years ago by repeated flooding of low-lying land by shallow seas in a hot climate, when Britain was at the latitude of what is now the

Sahara Desert. All in all, Britain's salt industries have a long and fascinating history and have made a significant contribution to our way of life, and even to life itself.

Britain has large deposits of common salt in Cheshire at Northwich, Middlewich, and Nantwich, and also at Droitwich in Worcestershire. 'Wich' means a salt works, 'Nant' means famous, and 'Droit' is dirty or muddy, and all these towns are in the Domesday Book (1086), though the names of Norwich and Ipswich refer to harbours. Where the salt beds are near the surface, the salt dissolves in rain and groundwater, and natural brine springs occur which have been known from ancient times.

Salt can be produced by evaporating brine or sea water. Sea water evaporation dates back at least 2,500 years ago along the south and east coasts, while in Worcestershire brine was evaporated. There were extensive trading links to the rest of Britain to make salt available for food preservation and for use in making cheese. Given the British climate, the work in these 'salterns' was seasonal and done by small groups of 'salters'. An alternative technique that did not depend on the weather was to boil sea water in large, flat earthenware pans, but this used a great deal of wood. The salt was scraped from the pans and put into pots or baskets to dry, the baskets making it easier to transport the salt on pack horses.

Salt-making in Roman Cheshire was a major industry: the army needed large amounts of leather, and salt for tanning became important. The salt workers may have been slaves or convicts guarded by soldiers, and their lives must have been bleak indeed. Anglo-Saxon salt-making occurred on a fairly large scale, with the development of laws and customs regulating the industry, such as the dues to be paid to the king or to the local lord. With the arrival of the Normans the Domesday Commissioners recorded what had existed before 1066, in 1066, and in 1086, and Droitwich was reported to be a major centre for salt, out-producing the whole of Cheshire both before and after 1066. In 1070 a rebellion against the Normans led to savage reprisals that devastated much of northern England including the three Cheshire wiches: only one building was left standing in Nantwich. The town suffered from raids from Wales, and a fire in 1583 burned for three weeks and required £30,000 for repairs—a very large sum at the time.

The salt industry matured in Tudor and Stuart times, and by 1485 the Cheshire industry was ancient and well-regulated. The traditional salt-pan method was still used, as it had been for several hundred years, Nantwich producing such fine-quality salt that the town dominated the Cheshire trade until the late seventeenth century. Salt boiling continued in Staffordshire and Worcestershire, and there were coastal salterns on the south coast and in the north-east, the latter using coal from the Tyneside pits.

Salt-making was called 'walling', from an Anglo-Saxon word meaning 'to boil'. The brine was drawn from the pits by men who went up and down ladders carrying it in leather buckets, though rag-and-chain pumps came into use in the mid seventeenth century. The pits were lined with timber to prevent the sides from collapsing. At the surface, the brine was poured into wooden gutters leading to the various salt houses, or wiches. The earthenware pans had now been replaced by lead, and each wich was only allowed a limited number of leads, typically six. At the wich, the brine was stored until the bell rang to show that walling was permitted under the detailed rules that governed everything, Nantwich having inspectors to enforce the regulations. It was important to ensure production in summer so that salt was available for the autumn slaughter and salting of meat. Salt was transported elsewhere in carts or on packhorses; Salterforth in Lancashire, a name that dates from the thirteenth century, means a ford used by salt merchants. In the seventeenth century, coal from south Lancashire started to replace wood for heating, which stimulated the growth of the Lancashire coal industry.

Living conditions in the salt towns must have been barely endurable. Tudor towns were in any case crowded and sanitation was casual, despite efforts to flush the streets from time to time. The smoke and pollution from the wiches made conditions even more dreadful, and became yet worse when coal came into use. Accidents, such as scalding from the boiling brine, must have occurred, and medical treatment was primitive, even if it could be afforded. Over and above all this was the grinding labour, especially for the men with their buckets in the brine pits, where accidents on the ladders must have been common.

A major event was the discovery of rock salt in 1670 by John Jackson while prospecting for coal at Halton, now part of Runcorn. This was such an important scientific discovery that it was reported in the *Philosophical Transactions of the Royal Society*. By 1697 four underground rock-salt mines were in production, and this abundant supply of good-quality salt provoked bitter struggles between the brine men and the rock men, eventually reducing the power of the established brine-based order. Within a very few years, salt works were established at Liverpool, and the expansion of the Cheshire salt industry, the rise of Liverpool as a major port, and the development of the south Lancashire coalfield, fed one upon the other.

The early salt mines exploited a thick bed of salt fairly near the surface. A shaft was sunk and a pipe with a fan provided limited ventilation, but little is known about the mining method. The rock salt would have been carried to the surface on the backs of men or pulled up by a horse whim. The pits were small, numerous, and short-lived—flooding was common because it was impossible to keep the water out. That left rock pit holes, which have disappeared due to subsidence caused by later deep mining. The rock pit industry prospered for more than 100 years, and large amounts of salt were transported by river to Merseyside where it was used in salt refineries and chemical works to enrich weaker brine.

In 1781 a lower and richer bed of rock salt was discovered near Northwich and deep mining began, gradually putting the shallow pits out of business. Each mine had two shafts to provide ventilation, but the big problem was water, so a third shaft was sunk to the level of the ground water with a pump to keep the water as far away as possible from the salt. The shafts were lined with iron rings, called tubbing, to keep water out, and the shaft tops were covered against rain and snow. Despite all these efforts, flooding still occurred and many mines were lost in that way. The brine from flooded mines, called bastard brine, was pumped out to feed the salt works, but such uncontrolled pumping led to subsidence which damaged houses in the surrounding towns. Compensation for subsidence was difficult to get as it was nigh impossible to prove who had caused it. All this pumping, as well as the hoisting of salt, was only feasible with the use of steam engines.

The salt beds are very thick, as much as 100 feet, but only the bottom 15 or 20 feet, which was of the best quality, was mined. The method is known as room-and-pillar, in the sense that a room was an area that was mined, but pillars of the same size were left as roof supports. Within a room, the top half of the salt to be mined was taken first, followed by the lower half, a process called roofing and benching. The salt was broken by blasting, with larger lumps broken by hand with a pick. The shot holes were initially drilled by hand with an iron bar, though compressed-air drills came into use in the 1880s. The broken salt was loaded by hand into small wagons or tubs running on rails and hauled by ponies to the shaft for hoisting. Gangs of miners would negotiate with the owner a price for mining a given section of salt. Boys over the age of 12 did jobs such as pony driving, but woe betide a boy who mistreated a pony. The poor ponies eventually grew too big to be taken back to the surface and ended their days below ground, unlike the ponies in the collieries, which were given an annual holiday on the surface and retired to green pastures. As far as we know, women and small children did not work underground in the salt mines, a contrast with coal-mining.

By 1900 only four deep mines were in production. The last one, at Northwich, closed in 1928, though in the late 1920s the Meadowbank mine at Winsford was reopened by ICI and is still worked as Britain's main source of salt for de-icing roads. The deep mines of Cheshire are, though, a pale shadow of those in Poland, where huge caverns have been mined out, complete with chapels and other votive carvings. The Russian salt mines were penal colonies.

The discovery of rock salt in 1670 had shown that the source of the brine at the surface was water penetrating the ground and dissolving the top layer of salt. That source could be tapped by sinking wells: dangerous, as the deep brine was under great pressure and, when the layer of hard marlstone—basically compressed mud—above the salt was penetrated, what oil drillers call a blowout could occur. On one occasion the drillers only survived because the steam engine was able to wind them up the shaft slightly faster than the brine was rising.

In medieval times the workers tended to live on the edges of the wich towns, the town centres being taken up by the salt works, though the workers' houses can only have been hovels. In the eighteenth century, with the growth of large salt firms such as that of the Marshalls (whom we will meet again in Chapter 12), the salt works moved to the edges of the towns and houses were built in town by the companies. These were rows of terrace houses—each row having a corner shop—that are still common all over the industrial north of England. These houses, now modernized, are often solid and well-built, but that was not always the case in Cheshire, where some houses, as well as schools, were built of 'bass' by using the cinders from the salt boiling. Crushed cinders were also used for road surfacing, which was dusty in summer and like inky mud in winter. Thus, although the salt industry brought much employment to Cheshire, and great wealth to some, it also brought serious pollution. The chimneys were often not high enough to carry away the smoke, and that was made even worse by noxious fumes from the smoking cinders from the boilings.

The salt industry expanded rapidly during the nineteenth century, mainly because Britain's population grew from 6.5 million in the mid eighteenth century to 21 million a hundred years later, but also because of the movement of people to the new centres of manufacturing in towns and cities, which then needed greater quantities of salt for food preservation and a host of other uses. The effect of these changes was near chaos in the salt industry, with periods of over-production and cut-throat competition. Several attempts were made to impose some sort of order by forming trade associations. This had little success until the formation of the Salt Union in 1888, which bought up 64 salt firms in Cheshire and elsewhere, although it never had a monopoly. The Union was never a complete success, but it did exploit a change in technology by importing a vacuum evaporation process from the USA. In this method, brine is placed in a large tank, the air is pumped out so that the water evaporates into this vacuum and the water vapour can be extracted to leave the salt behind. This requires much less fuel than pan-boiling, though the latter continued for many years as it could produce many different grades of salt, depending on the size of the pan and the length of heating. In contrast, the vacuum system at first made only fine salt crystals, and it was not until the 1930s that it was developed to meet all demands.

A social change in parallel with the growth of the industry in the nineteenth century was the formation of workers' friendly societies, aimed at supporting members in distress but also at convivial entertainment through tea parties, brass bands, and the like. The pub also played a part. Trade unions evolved naturally in response to strikes and disputes. One of these, in 1888, lasted for two and a half months, but the employers eventually accepted the workers' demands. Major trouble ensued in 1892 when the watermen, who transported salt on barges to the chemical works on Merseyside, struck for shorter hours—they were expected to work for 30 or 40 hours at a stretch. The salt workers' union supported them and rioting occurred that was so bad that the police could not control it. The rioters were eventually persuaded to disperse, though not before troops had arrived from Manchester. A settlement was reached and the excessive hours worked by the watermen were reduced. There was much distress in Cheshire in 1893 when a lockout in the coalfields cut off the source of fuel. Fortunately, by the end of the century the firm of Brunner Mond dominated employment in Northwich, though they were chemical manufacturers rather than salt makers. They established much better working relationships which were carried over when Brunner Mond became one of the founding members of ICI. In fact, the next strike in the mid-Cheshire chemical works did not occur until 1960.

Today British Salt, a subsidiary of an American firm, extracts brine from strata 600 feet below the surface. Water under high pressure is pumped in to dissolve the salt and the resulting brine is pumped about 3 miles through underground pipes to Middlewich, though the brine extraction is carefully controlled to prevent subsidence, and non-toxic waste is pumped back below ground. An advanced six-stage vacuum evaporation technique is used to minimize the energy requirements and it produces 800,000 tons a year of pure salt. This is a very far cry from the grinding labour and terrible pollution of only 150 years ago.

We'll now move to Staithes in Cleveland on the north-east coast, and visit Boulby Potash Mine, the surface works of which were designed by a prominent architect to harmonize with this very beautiful part of Britain. The mine dates from the 1960s when Britain was heavily dependent on imported potash for fertilizer. Potash is the term for chemicals based on potassium. ICI decided to develop Cleveland's potash reserves. An important one is potassium nitrate, which is the basis for fertilizers and explosives and which can be manufactured from sylvite (potassium chloride).

The ore is called sylvinite, which consists of about 50 per cent sodium chloride, 40 per cent potassium chloride, and some clay impurities. The seam, which averages about 20 feet thick, is between 4,000 and 5,000 feet below the surface and is beneath a layer of sandstone that contains water under high pressure. To sink shafts through these dangerous strata special techniques involving freezing and cement injection were used, and the shafts are fitted with watertight linings. An added difficulty is that the rock temperature is over 40°C, so a high-capacity ventilation system is needed to maintain acceptable working conditions. The strata also contain pockets of gas under very high pressure, which can cause outbursts of rock when the machines cut into them.

Boulby is now one of the biggest mines in Britain, employing about 800 people and producing 3.5 million tons of ore a year, yielding more than a million tons of potash and half a million tons of salt. The potash goes for fertilizer and the salt for road de-icing. It is a tribute to the care of the mine planners and the architect that one can drive through the area and hardly know that the mine is there. A fascinating aspect of Boulby is that the great depth makes it suitable for scientific experiments by the Institute of Underground Sciences to detect so-called 'dark matter'—an enigmatic form of matter that does not reflect light—and extensive laboratories have been built underground.

The raw material for explosives, nitrate, had traditionally been imported from Chile, but during the First World War the German submarine blockade threatened that supply, so the British government bought land at Billingham on Teesside and established a factory to make synthetic nitrogen products. After the war, the site became part of ICI. It turned out that the factory was on top of large deposits of anhydrite, and mining commenced in 1926, although to prevent subsidence in the factory above the mine, only about half the available anhydrite was extracted. At the working face 60 holes were drilled, starting with four in the centre of the tunnel, converging to create a pyramid effect. More holes were drilled further out from the centre, but at progressively shallower angles, until the holes at the edges eventually went 10 feet straight into the ore. On the next shift the holes were charged with 3 pounds of dynamite each and were set off in a sequence of short delays so that the pyramid was blown out first, leaving space for the other charges to do their work. The third shift in the cycle used a machine called a gathering-arm loader to collect the 300 tons of ore broken by each blast. This can be visualized by imagining yourself on the beach scraping up sand for a castle but using alternate arms, so that as one arm scrapes the sand forward the other reaches out for more. The mechanical loader works in the same way but can handle about 100 tons an hour. The mine closed in 1971 but one civilized feature of it was the underground messroom, fully tiled, and with washbasins and lavatories.

Barrowmouth in Cumbria was also an important source of anhydrite; 5 million tons of it lying beneath an old coal mine. The anhydrite mine had to avoid gas and water in the old colliery and, almost uniquely in non-coal mines, the anhydrite miners had to work to coal-mine standards, such as not being allowed to smoke. The mine opened in 1955 and was highly mechanized, allowing 60–80 men to produce 43,000 tons a year. Each ton of anhydrite, combined with phosphate shipped from Morocco, produced a ton of cement and a ton of sulphuric acid. Faced with a global surplus of sulphur, the mine was forced to close in 1976, and was sealed in 1984. It is interesting to note that the workforce was easily recruited from men made redundant from other Cumbrian mines who eagerly abandoned factory jobs in order to go back to mining.

Gyspum, hydrated calcium sulphate, is a very important and valuable salt with many uses. When gypsum is heated to 1700°C it loses about 75 per cent of its combined water and can be ground to a fine, dry, powder—plaster of Paris. However, when the powder is mixed with water, it rapidly becomes very hard, and can be used for splinting broken bones and for making all manner of plasters, including for delicate stucco ceilings. It has the added benefit of being fire-resistant. Other uses of gypsum are in the manufacture of cement, paints, and fertilizers, and for polishing glass. However, while most salts are mixed with some sand or mud from the sea in which the evaporite was deposited, gypsum is sometimes found in a very pure state, in which case it is called alabaster, which can be carved into, statues, ornaments, and architectural decorations. Since gypsum is somewhat soluble in water, alabaster is not recommended for garden ornaments.

Gypsum and alabaster occur in Cumbria and were exploited by local people and companies, usually in quarries or pits: the mineral was called 'silver in the woods'. Locally it was used to whiten the edges of doorsteps, but the demand for plaster increased as homes, especially grand houses, became more prosperous and included lavish plasterwork. A significant step was the invention in 1885 of

Robinson's Cement, which was used in the Houses of Parliament and as far away as Capetown. The Cumbrian gypsum industry prospered and lasted: for instance, Knothill alabaster pit worked from 1685 to 1930 and the Long Meg mine in the Eden valley worked a seam 18 feet thick, 14 feet of which was gypsum, from 1870 to 1976.

There are large reserves of gypsum in Leicestershire and Nottinghamshire, and British Plaster Boards operates underground mines, the largest being at Barrow in Leicestershire, and some quarries. The underground mines use the room-and-pillar method and extract as much as 75 per cent of the available ore, with no subsidence and without significant waste. Current output of natural gypsum and anhydrite is some 1.7 million tons per year, but a further 1.4 million of synthetic gypsum is produced at power stations, and some is imported. Of this total, 3 million tons is used in plaster and plasterboard, and 0.6 million in cement.

This industry, now more than 300 years old, still flourishes, in contrast with other parts of Britain's mining heritage which seem to have had their day. We will return to this question in Chapter 14.

Alum was very important for dying fabrics in the Middle Ages, because the aluminium oxide that it contains coated textile fibres, enabling vividly coloured vegetable dyes to be used reliably, something that had been known in ancient Egypt as early as 2000 BC. Supplies from overseas were expensive, but shales on the cliffs along the coast from Whitby proved to be a productive source of alum. From about 1604 a large industry eventually developed, with 24 alum quarries running from the coast and inland as far as Guisborough, though the coastal quarries made bulk transport much easier. Alum was also produced in Scotland and in Dorset, though on the Isle of Wight the place name 'Alum Bay' denotes a failed attempt to find this valuable mineral.

The alum from the cliffs was roasted, soaked in water, and boiled. The liquor was crystallized by adding potassium sulphate from seaweed or aluminium sulphate from human urine. Of course, all this chemistry was only worked out centuries later, and we can but admire the people who developed it by trial and error, or perhaps by industrial espionage from the Italian alum industry. Horses and carts brought coal from Durham over the rocks from flat-bottomed boats that could be beached safely for unloading and afterwards floated off to take the processed alum to market.

The alum industry took decades to become profitable: by 1612 it was producing about 600 tons per year which cost £30 per ton but sold at £23. However, the textile industry was so important that the Crown took over the alum quarries and had invested £50,000 by 1615, though it was not until the 1640s that a return was achieved, after which alum was a nice source of royal income. Money was also invested by local magnates, and alum was one of Britain's earliest capitalist industries.

Working conditions were abysmal. Sulphurous fumes filled the air and left a scum on hedges and trees. Working on the cliffs in a North Sea gale was very dangerous and it was not unknown for a boy to be blown into the sea. The trade was interrupted by wars— Boulby quarry bought a cannon to deter raiders. Wages were pitiful: even as late as 1788, pickmen in the quarries earned two shillings a day, but it was as high as four shillings in the alum house where the processing happened. By 1846 the pickmen were on 2/6 (12.5 pence), though the alum workers were still no better paid. The best that can be said is that their cottages usually had large vegetable plots.

The workers also suffered because quarries tended to close when trade was slack or the weather too bad for working. Some operations were more successful and Sandsend ran from 1610 to 1870, when the industry died. It was killed off when a process of treating shale from coal mines with sulphuric acid was invented in 1846. Nowadays, dyes are made synthetically and the main use of alum is in styptic pencils for minor cuts.

The legacy of the alum industry can be seen in the cliff-side quarries along the coast from Whitby northwards.

9

Iron and Steel: The Sinews of Industry

About 1000 BC the Bronze Age passed into the Iron Age when the mining and smelting of iron became feasible. Iron mining started in Britain in the first millennium BC in places as far apart as Shetland, Sussex, and Somerset, and the first shoots of what became modern Britain began to flourish. Britain is blessed with large reserves of iron ore, so there was considerable production from ancient times. As well as the mining and quarrying of iron ore, we'll also have to look at steel and some of the metals, such as manganese and tungsten, that make steel so versatile and which are also found in Britain.

After aluminium, iron is the most common metal, averaging about 4.6 per cent of the earth's crust. In Britain a common iron ore is ironstone, iron carbonate (48 per cent iron). Ironstone is common in the coal measures and is the result of precipitation of the mineral in lakes and seas. Other ironstones were formed when iron carbonate replaced calcium carbonate in limestone; these occur in Cleveland and the English Midlands. Another, and very beautiful, ore is haematite, a form of iron oxide (70 per cent iron), which has been mined in Cumbria and in the Forest of Dean.

For a metal, iron is soft and easily worked, which a blacksmith does by heating and hammering, a skill that it is a delight to watch. The beautifully ornate wrought-iron gates of palaces and mansions are one outcome of this craft. Another is very solid bars that clearly say 'keep out' or, at prisons, 'stay in'. For more demanding roles, steel is iron with a small percentage of carbon. Varying the proportion of carbon produces steel with different properties, such as the ability to take a sharp edge. Iron has many uses in its own right, such as horseshoes, and steel can be made into anything from surgical scalpels to supertankers, so these metals have been essential to our well-being and prosperity.

There are numerous varieties of steel, a very common one being *stainless steel*, used in cutlery and often called 18/8 as it contains 18 per cent chromium and 8 per cent nickel. Changing the proportions of these, or replacing them with manganese, produces steel for special purposes such as gun barrels and armour plate. Molybdenum steel is super-hard and can cut other steels. Some 'steels' contain very little iron but large amounts of nickel or cobalt and are used under high temperature conditions, such as in jet engines. The maraging steels are various alloys of iron, cobalt, molybdenum, and titanium. Their extreme toughness makes them good for missile and rocket skins, engine crankshafts, and golf club heads. Maraging steel is used to make centrifuges for enriching uranium, so its use by certain nations is carefully monitored.

A very important source of iron was in the Weald, between the North and South Downs and covering most of Kent and Sussex, although the High Weald, which stretches between Hastings and Horsham, was the main mining area. The Weald was, and still is, heavily wooded (the name comes from German 'wald', or forest), which supported much charcoal burning to meet the needs of the iron makers. The ore mineral is ironstone, or iron carbonate, laid down between 140 and 120 million years ago, the ore occurring as nodules or beds in the clays of the Weald. Iron mining was started in about 100 BC by the Belgic peoples, who traded with their fellows in Roman Gaul. Before we look at the history of Wealden iron it is necessary to say something about the mining methods and the process of converting ore into metal.

The Weald has many deep valleys which made prospecting fairly easy, and once a bed of nodules had been discovered the ore was extracted. The pre-Roman miners worked from pits or small quarries, but from medieval and later times shafts were sunk, though mining was mainly a summer occupation. Once extracted from the ground the ore was roasted to convert iron carbonate into iron oxide though the early miners were not aware of that chemistry and worked from empirical knowledge. The roasted ore was smelted, using charcoal, in a 'bloomery', or furnace, that was little more than a clay oven. To raise the temperature so that iron would melt, air was forced into the furnace using leather bellows, the air passing through tuyères, or air pipes, made of clay. Two bellows, operating alternately, gave a steady blast, and much of the gangue was removed from the ore as slag, though the slag still contained some iron.

The product was a bloom weighing only a few kilograms, but that was a spongy mass of iron and slag, so the next stage was to heat it to 1200°C, with temperature judged by eye, and hammer it. This was repeated as often as needed, to drive out the slag and thereby produce iron that a blacksmith could make into ploughshares, fittings for farm carts, or whatever was required for local use or trade. In the Middle Ages water power may have been used to drive the heavy hammers and the bellows. From the late sixteenth century, local streams were dammed to form hammer ponds to ensure a good water supply, and waterwheels became a common feature of the Wealden iron industry. The old hammer ponds at Abinger Hammer in Surrey are now used for growing watercress.

Historically, iron was mined in the Weald in two distinct periods. The first of these started with the pre-Roman people, and continued until the end of the Roman occupation: pottery and charcoal from the Roman period have been identified by the Wealden Iron Research Group. Output seems to have been in the region of 700 tons per annum between AD 100 and AD 250, but thereafter it dropped sharply, possibly because Saxon raids forced a move to the Forest of Dean. The post-Roman record is vague until medieval times, when there were large orders from the Crown for steel arrow-heads—literally millions of them—and other equipment for the wars with France. In the late thirteenth century, the Sheriff of Sussex was ordered to supply 30,000 horseshoes for the royal army.

The second period started about 1490 when bloomeries were gradually replaced by blast furnaces though the change-over took many decades to complete. The blast furnace operated at a higher temperature than the bloomery and, because it was a more substantial, brick-built structure, lasted longer. It used charcoal and ore, but sometimes with the addition of limestone as a flux, which made the molten iron more fluid and removed impurities in the form of slag. More ore, charcoal, and iron-rich slag could be fed into the top of the furnace so it could be operated for long periods, called campaigns, thus producing much larger quantities of iron, which was run into sand moulds to form ingots resembling a sow with a litter of pigs; hence pig iron. Pig iron contained a good deal of carbon and was fairly brittle, so it was heated and hammered again in a 'finery'. This reduced the carbon content from 3 per cent or more to 0.1 per cent, and it was then called bar iron, though the process could be stopped at about 0.3 per cent to give steel, and this was all done by eye and experience.

Bar iron can be converted by a skilled smith into just about any wrought-iron artefact; decorative objects, such as beautiful firebacks, were made by re-melting bar iron and casting the desired object, but a very important part of Wealden output was cast-iron cannon, the invention of which was a major achievement. Prior to that, cannon had been made from wrought-iron bars formed into a cylinder and held in place by stout iron hoops, or from gunmetal, a type of bronze, which was easy to cast without the flaws or cracks that it is crucial to avoid in a cannon. Bronze guns had to be allowed to cool after a few shots and were very costly. The innovation for iron guns was to cast flawless iron, which took much trial and error but had the benefit of being a fraction of the cost of a bronze piece. Recent research has shown that the iron cannon were standardized with accurately manufactured shot, some of which were used in the defeat of the Spanish Armada in 1588. In a sense, mining played a small part in shaping Britain as a mainly Protestant, not Catholic, country.

Despite these achievements spanning nearly 2,000 years, by the early eighteenth century the Weald could no longer compete for general-purpose iron with the large quantities and low costs of iron from blast furnaces in the Midlands and Scotland. However, the area's ironmasters retained a near monopoly of cannon supply until later in the century, helped by a spurt of demand from the Seven Years' War of 1756–63.

The last furnace in the Weald closed in 1813. The industry's legacy in the landscape is the depressions in the ground left by many thousands of mine pits, and the remains of working sites. Perhaps a more important inheritance is the woods of Sussex and Kent, which are the consequence of careful management, over centuries, in order to maintain a supply of charcoal, which is made from wood coppiced—essentially harvested sustainably—at a small size and quite a young age. Thousands of acres of coppice were needed to supply all the furnaces. However, carpenters wanted larger and older wood for building houses and farm wagons, and boat-builders had their own requirements, often from different species of tree. A wheelwright used oak for the spokes, ash to make the rim, and elm for the hub, but the best charcoal is made from hazel. The potential problem is that coppiced wood grows again within a few years, while an oak tree big enough for roof beams will take decades to mature. The inevitable arguments about who got the wood he needed were solved by management of the woodlands so that large trees still had space to grow, with coppice grown below them. That was fine in principle, but landowners found that they could get a faster return by selling coppice wood for hop poles and fences before it was old enough for charcoal-making. The resultant rise in the price of charcoal was another blow to Wealden iron-making, but the industry has made us a bequest of beautiful and varied woodland.

The Forest of Dean to the west of the River Severn has a long history of iron and coal mining. Slag has been found in an iron-age hill fort, while the Romans have left extensive reminders of their work. Iron mining revived after the early Middle Ages; the monks of Flaxley Abbey, for example, were involved in the iron trade. By the reign of Edward I there were some 60 mines, and in 1296 miners from the Forest were 'recruited' to go to Berwick-on-Tweed to undermine the town's defences during a war with the Scots, though what these men and their accompanying wives thought of this journey defies imagination.

When they arrived their job was to undermine the walls, protected by a wooden roof against the hostility of the defenders, and supported with massive timbers. When the hole was deemed to be big enough, the cavity was packed with brushwood and soaked with lamp oil, the mixture was set ablaze, and the section of wall collapsed. In the resultant sack of Berwick, some 7,000 of its citizens were killed, though the destruction of a captured city and the slaughter of its inhabitants was 'normal' for the times. The king showed his gratitude for the victory by granting free mining rights as the 'Laws and Customs of the Miners in the Forest of Dean', which continue to apply to coal mining in the forest. The rights to mine in an area called a 'gale' were granted by the king's officer, the Gaveller. That office still exists and the Gaveller maintains the plans of abandoned mines.

In the seventeenth century, mining improved, thanks to better furnaces, but Parliament became concerned about the availability of oak for shipbuilding during the 20 years of the Dutch Wars and placed restrictions on the industry. It was soon in a poor way, and it was not until the mid nineteenth century that the development of efficient blast furnaces revived demand, though powerful pumps had to be installed as the mines went deeper. The revival was transient, and most mines had closed by about 1900 due to diminishing reserves, high costs, and competition from Spain. During the First World War a few mines re-opened, but most closed by 1926. One mine stayed open and was controlled by the Ministry of Supply during the Second World War, finally shutting in 1946 after more iron had been put in for rails and roof supports than had been taken out.

Throughout medieval times iron mining continued in Britain, often managed by the monasteries and abbeys, which needed a source of income to support their communities and charity. For instance, the brethren of Kirkstall Priory in Leeds ran a factory in Bradford which produced iron wire. Rievaulx near Helmsley in North York-shire is now a delectable spot with a small village, a verdant valley, and the ruins of the Abbey. The Cistercian order founded the Abbey in 1131, but within 20 years it had 140 monks and 500 lay brothers. The latter were celibate and had to attend some of the

Abbey services, but their main role was to work the farms and iron mines. Being a lay brother was, despite the celibacy, an attractive proposition, as it guaranteed food, shelter, and care in later life. Charcoal was made in the surrounding woods and the valley was a hive of industry. The trade was so productive that the iron was exported via the Humber through what is now Hull. When the monasteries were dissolved by Henry VIII in 1535, the property was obtained by the Duke of Rutland, who carried on the trade in what is now Rievaulx village, but production ended in 1647 as a consequence of the Civil War.

Iron has been mined from haematite ores in Cumbria for probably a thousand years. The Norse settlers are likely to have mined iron for farm implements and weapons, though later medieval workings have obliterated the traces of earlier mining. Iron ore was mined in the Lake District at Grasmere and elsewhere, but the main ore deposits are in the coastal areas around Barrow-in-Furness, Ulverston, and north towards Whitehaven. The coastal mines seem to have been started by about AD 1200 by the monks of Furness Abbey, initially using open-cast methods, though later mines went deep and sometimes encountered huge ore bodies. The peak years were in the 1870s and 1880s, after which there was a steady decline until the last big mine closed in 1946. Mining ceased altogether in 1968.

Some of the Cumbrian mines were major operations. For instance, the Hodbarrow mine at Millom was very successful, in terms of not only the quantity of output but also the quality of the haematite, which supplied a local ironworks. Before 1855 there had not been much success, but in 1868 a huge ore body was discovered after boreholes had been used to explore the ground. For 50 years the deposit was mined by top-slicing in which the successive layers of ore were removed, allowing the overlying limestone to collapse, but this caused problems due to subsidence and because of the proximity of the sea, so the method was changed to bottom-slicing, with a slurry of water and sand pumped in to stabilize the ore body. The Hodbarrow mine ceased production in 1967 after 112 years, and in 1968 the ironworks closed.

In what seems like another of the amazing coincidences we looked at in Chapter 1, the Cumbrian coast also had coal, limestone, and excellent harbours at Barrow, so a steelworks and a big shipbuilding industry prospered. Submarines and warships are still built at Barrow for the Royal Navy. The complete vessel is constructed in a gigantic shed and then wheeled out on a huge trolley and lowered into the water. The Dock Museum at Barrow has records of ships built there, the models that were made to show the potential owner what the vessel would look like, the story of the mining, and the lives of the people.

The coal measures of the Carboniferous system usually contain bands of ironstone which gave rise to great iron- and steel-making industries in places such as Rotherham, Sheffield, Ebbw Vale, Port Talbot, and the famous Carron Ironworks at Falkirk. South Wales had everything: good-quality coking coal; limestone as the flux for steel-making; clay bricks for the workers' housing; refractory clay for furnace linings; large resources of limonite, a form of iron oxide; and a large and skilled workforce. The iron ore was mainly quarried from outcrops, and large tips of waste were cascaded down the hillsides to form patchworks; just outside Brynmawr the tips came close to engulfing the edge of the town. With such resources, the Ebbw Vale steelworks started in 1789, and it finally closed in 2006. The Port Talbot operation still continues, now under Indian ownership.

Cleveland and the eastern part of North Yorkshire supported a very large ironstone-mining industry which, from the middle of the nineteenth century, provided the raw material for great steelworks at Middlesbrough, Darlington, and Hartlepool. Iron, and later steel, rails were rolled at Middlesbrough for railways in Britain and much of the world. The iconic Sydney Harbour bridge was designed and built in 1932 by Dorman Long and Co. Ltd of Middlesbrough, and the steel was shipped out to Australia. Generations of engineers used the company's *Safe Load Tables* to calculate the load that could be placed on various sizes and types of steel girder in different circumstances. The steel produced also supported major shipyards on Teesside and the Durham coast, all now sadly gone.

There is evidence that iron was mined in Cleveland in prehistoric times, perhaps as early as 700 BC, and in medieval days iron was smelted by Guisborough Priory. Modern mining started in 1847, just in time for the proliferation of the railways, at Skinningrove near Saltburn. The first mines were small until, in 1850, a major deposit was discovered at Eston, conveniently situated a few miles east of Middlesbrough. Eston became the biggest drift mine in Cleveland (access was by adits), but eventually about 80 mines were worked in a large area from Eston as far south as Rosedale. Many railway lines were built running from Saltburn to the mining villages.

The Cleveland mines exploited a type of ironstone in which iron carbonate has replaced calcium carbonate in limestone. Large lumps of ore can be found on the rocky beaches, and are instantly recognizable by their rusty-brown colour and great weight; I use one as a book-end. Inland, the seams outcrop on hillsides and beaches where the early miners could easily exploit them, but the rich deposits are in seams at depths of about 230 feet, the main seam being about 13 feet thick, below which are two more seams of poorer quality ore. The deeper mines were worked by shafts or from adits, with ventilation and hoisting power from steam engines or electricity, as they became available. Ore was taken by sea to Tyneside and Middlesbrough, sometimes via ships that went onto the beach at high tide, were loaded, and then were refloated, as was done for the alum quarries, or from Whitby harbour. The developing railways also handled large tonnages.

The mines were operated on a room-and-pillar system, extracting perhaps half of the available ore. In all, some 350 million tons came from Cleveland's mines and quarries. The main seam was, unusually for a metal mine, gassy, and gas explosions had killed 11 men and injured 123 before it was found that the ironstone was impregnated with oil from the overlying shales which contain jet, a decorative form of lignite mined near Whitby (see Chapter 10). The worst accident was in 1953 when eight men were killed and others badly burned.

Although the work was hard and, as we have seen, not without its risks, the small mining settlements, which are the origins of the area's numerous small villages, had a strong sense of community, born of shared identity in the work of the mine. The houses were usually built by the mining company in terraces of 'two up, two down' structures, often with a garden or allotment at the rear, but the company also built schools, hospitals, and chapels. As in many mining communities, religion had a major part to play, but secular hobbies such as whippet racing, homing pigeons, and football were popular, and many pits had their own silver band. It does not do to idealize the life, as it was hard and all too often short, though it must have been far better than the slums and pollution of Middlesbrough. Sadly, this community identity largely died when Cleveland mining was affected by foreign imports. The last mine closed in January 1964.

The English Midlands were home to major iron mines, mainly in Lincolnshire. Pieces of ore, with their distinctive colour and weight, can often be seen in the gravel of drives and car parks. The ore deposits were widely spread across the county, one running from Lincoln as far south as Grantham and beyond, while others were in the vicinity of Scunthorpe. Most workings were open-cast quarries, but about 16 mines were underground. As there was no tradition of mine-work in Lincolnshire, men were imported from places such as Cornwall and Derbyshire, but local men soon saw that the mine might be easier work, and better paid, than long hours of labour in the fields. The first mine opened in 1868 and closed in 1885 after 500,000 tons of ironstone had been extracted. Thereafter, mines and quarries came and went until the last one closed in 1968, though there are still large reserves of ironstone.

Much of the product of these mines went to feed steelworks in Ilkeston, now closed, and Scunthorpe, where the steel industry is still a major employer. The industry there has contracted considerably, from once employing 27,000 people to the current 4,500, but it is still the major integrated steelworks in Britain.

Iron was mined in many parts of Britain, and we have no space for them all, but the island of Raasay in the Inner Hebrides is particularly interesting. Its complex geology includes ironstone beds, which were worked from 1912. In the First World War many of the island's men were serving in the army and were replaced by German prisoners of war. That probably violated the Hague Convention, which prohibits the employment of prisoners on war work, but a way round it was found by some clever legal manoeuvres. The mine closed in 1922, although about 10 million tons of ore remain.

Manganese is an important metal used, as we have seen, to make hard steels. Another important use is in manganese bronze, which is a complicated alloy containing 60 per cent copper, 25 per cent zinc, with the rest consisting of aluminium, iron, and manganese. Its high strength is effective for large engines and bearings for shafts running under heavy load at low speeds, such as the propeller shafts for ships. Manganese also has uses in the glass and chemical industries. The industry was never very large, but deserves a mention for its importance in engineering (the Manganese Bronze Holdings company is not engaged in mining but builds London taxis).

Britain's main manganese resources were in North Wales in the Lleyn Peninsula and the hills inland from Harlech in Merioneth, which geologists call the Harlech Dome. This is an area of about 100 square miles of harsh terrain with a less than gentle climate. Manganese ore was being mined on a small scale in Merioneth in the 1820s and was sent to Glasgow to make bleaching powder for the textile industry—many minerals seem to have almost limitless and unexpected uses—but towards the end of the century demand for the metal increased rapidly because of its use in the steel industry. From its tiny beginnings, the manganese industry expanded considerably to the extent that about 17 mines were working in Merioneth by 1890, together with a few more in Lleyn. After that, however, mining declined, partly because of poor ore quality and the physical difficulties in this landscape, but also because of foreign competition. Production revived a little during the First World War, but thereafter the industry declined until the last mine closed in 1928.

The main ore mineral is pyrolusite, or manganese dioxide, and the ore beds are only about 12 to 18 inches thick with a roof of bluestone, an igneous rock harder than granite. Initially the ore was worked from outcrops and from adits driven into the ore bed. From the adit, working faces were developed, but the bluestone had to be taken out first to give working height, after which the pyrolusite was lifted, nearly always with hand tools, though compressed-air drills and explosive were used when mines were re-opened in the First World War. The roof was supported by wooden props or packs of bluestone, the rest of the gangue being tipped outside. Transport in the mine was by barrow or sledge, and on the surface by horse and wagon on 'roads' that were only cart tracks, though aerial ropeways were used to a small extent. The miners often lived in barracks, or hostels, as was common in remote and inaccessible areas such as the Pennines. The pay was in the region of the equivalent of £1.75 per square yard extracted (which had to be split between three men and a boy, and also had to pay for explosives). There are certainly reserves of manganese, but it is unlikely that mining will ever resume.

Tungsten is another versatile metal much used in making very hard steels, such as armour plate and tool steels, though the process was not discovered until the late nineteenth century. It is also used in electric light filaments. The ores of tungsten are found in granite areas. Castle-an-Dinas mine near St Austell in Cornwall was the only mine in Britain that produced nothing but tungsten ores. Demand peaked in wartime, but shortage of capital and supplies meant that it was worked with hand drills and steam power from 1917 up to its closure in 1958. A bigger source was Hemerdon, just outside Plymouth. Tungsten was found there in 1867 and tin in 1916. An open-pit mine was worked during both world wars, but it has not been operated since 1944. It is estimated that there are still 40 million tonnes of tungsten and tin ores, and it is possible that mining will resume. Outside the south-west, Carrock Fell in the Lake District is significant. The area is very rich in minerals: lead, arsenic, iron, and tungsten have all been mined there, and Carrock Mine in Grainsgill Beck was the largest tungsten mine outside Devon and Cornwall. Mining started in 1854 and continued intermittently, but was intensive during both world wars and the Korean War of the 1950s. The mine closed in the 1980s and the site was bulldozed a few years later.

10

The Riches of the Earth: Britain's Gold, Silver, and Precious Stones

Mankind has lusted after gold, silver, and precious stones since time immemorial for adornment and as status symbols. Their rarity, durability, and desirability have ensured their continuing value. Gold does not decay, so it was once used as coinage, and today it is stored by nations as a guarantee of their currency. The Bank of England holds gold, which was once kept in its vaults. Until as late as the 1970s a detachment from a Guards regiment was stationed at the bank overnight. Individuals also hold gold, often in the form of classic coins, such as the golden guinea, or the Maria Theresa gold dollar. The price of gold is not stable, but in times of economic recession, such as the crash of 2008, it can be relied upon to fulfil its ancient role as a thing of fundamental value.

In India a bride is adorned as lavishly as possible to demonstrate family status, so about 25 per cent of world gold production is now sold there. The amount increases when the price is low, but the gold shops are always a sight to behold. Gold's other uses are in dentistry and high-quality electronics. Silver has many uses, but jewellery is an obvious one. In fishing villages on the Dutch coast, one can still see old ladies, clad in black, with elaborate silver pins in their head-dresses. The pins showed the love and success of their fishermen husbands, but they were also a buffer against bad times.

Much of the world's gold now comes from places such as South Africa, Australia, Russia, and the USA and while Britain has nothing to approach such riches, it has produced surprising amounts since ancient times. As we have seen, the tin-streamers in Cornwall would occasionally find flakes of gold, store them in a hollow quill, and after they had accumulated enough, sell it to goldsmiths. The Romans mined gold in Wales and it was mined on a small scale at Tyndrum in Perthshire. That mine may re-open if the price of gold remains high, but we will concentrate here on the main Welsh gold fields.

The oldest gold mine in Wales is at Dolaucothi near the village of Pumsaint, between Llanwrda and Lampeter. It was certainly worked by the Romans, possibly with convict labour, but is likely to date from earlier times. The Roman workings involved trenches, large pits, underground stopes which eventually went about 140 feet below the present ground surface, and placer deposits below the outcrop of the gold ore. The surface works alone produced about 500,000 tons of ore, yielding an estimated 830 kg of gold. Dolaucothi was probably the only gold mine in Roman Britain, but it is not clear how long the Romans mined there. In AD 105 the Roman Emperor Trajan conquered Dacia, modern Romania, the booty including much gold and silver, so Dolaucothi probably became unimportant to the empire.

Archaeological study by Lampeter University and research by the former Mining Department at Cardiff University has shed much light on the Roman mining methods. The ore was extracted by fire-setting—heating the rock—and hushing, which we met in Chapter 5 on lead. The lumps of ore would be smashed with hammers and ground to small sizes and the gold recovered by washing over tables. That required a lot of water, which came from the Cothi river via leats extending about 7 miles over careful gradients and capable of providing some 2.5 million gallons a day.

By the early nineteenth century the mines had long been abandoned. It was known that they were of Roman origin but no one knew what had been mined, and it was not until 1844 that it was realized that gold had been won from Dolaucothi. Sir Henry de la Beche, the first director of the British Geological Survey, assayed the amount of gold in ore samples, the results suggesting that it might be a commercial prospect to re-open the mine.

From de la Beche's day until 1939 the mine was worked on and off by a succession of companies, including the grandly named Roman Deep Holdings. Profits were uncertain, though £11,000 worth of gold was produced in 1938. One reason for the spasmodic profits was that conditions were difficult underground because the hanging wall was what miners call 'heavy-backed', meaning that is was soft and liable to collapse unless strongly supported. The ore also became more difficult to process as the gold was trapped in iron pyrites (iron sulphide), and arsenical pyrites (iron-arsenic sulphide), which meant that complex chemical processes were required to extract the gold. No British refining company would process the relatively small amounts that Dolaucothi produced, so it was refined in Hamburg, though that ceased with the outbreak of war in 1939 when the mine closed.

In the 1960s Dolaucothi became a National Trust property which, until 1998, was used as a training mine for Cardiff University's students. There is still gold ore in the ground, and the mine might have some future potential as a combination of a small-scale working mine and a tourist attraction.

The other gold area in Wales is the Dolgellau mines in Merioneth, which roughly follow the Mawddach river inland from Barmouth to Dolgellau and then northwards along the A470. In this small area there were four fairly large gold mines and about 14 smaller operations, most of the latter employing very few men. This part of Wales is richly mineralized: the famous slate quarries at Blaenau Ffestiniog are a little further north, manganese occurs in the hills behind Harlech, and the Mawddach valley has seen copper and lead/zinc mining; in 1836 it was prospecting for copper that led to the discovery of gold at Clogau. Clogau operated successfully, but the big discovery was made in 1867 at Gwynfynydd, and that mine became prosperous under the subsequent management of William Pritchard Morgan.

Morgan was a remarkable man who had emigrated to Australia in 1844. He became a farmer, but in his spare time qualified as a lawyer, an occupation in which he became prominent and successful, acquiring a good knowledge of mining in the process. He came back to Wales in 1883 a rich man, and soon invested in the Merioneth gold mines. Pockets of rich ore were found in 1887, some samples assaying at the very high values of 20–40 per cent gold; more like quartz held together by gold, rather than quartz containing the metal. Even bigger discoveries were made at Clogau in 1900 and in 1904, but such bonanzas were unpredictable and were interspersed with 'barrascas', or hard times, when there was little alternative employment in this mainly agricultural area. The mines struggled on under changing managements, but Clogau was shut in 1995, even as a tourist attraction, and Gwynfynydd finally closed in 1998 because of geological difficulties and pollution control legislation

PLATE 15 Penrhyn Castle, showing the vast wealth accumulated by a quarry owner.

PLATE 16 A Pennine miner's small farm.
The plant on the roof is a leek—grown for good luck.

PLATE 17 Iron-making at Fernhurst, West Sussex, an artist's imaginative reconstruction. In reality the cannon were cast and bored elsewhere.

PLATE 18 The legacy of iron-quarrying. Patchworks at Brynmawr, South Wales in 1947. The area is now houses.

PLATE 19 A 'flat' at Gunnersby iron mine, Cumbria.
The roof is supported with stone packed into a wooden frame.

PLATE 20 Hydraulically powered underground winding engine giving
access to a deeper shaft.

PLATE 21 Adit driven by Roman miners at Dolaucothi gold mine, Carmarthenshire

PLATE 22 Wentworth Woodhouse Hall, a coal-owner's mansion. The facade is twice as wide as that of Buckingham Palace.

PLATE 23 Thomas Williams, the 'Copper King', painted in 1787 by Sir Thomas Lawrence, one of the most eminent and expensive portrait painters of the day.

PLATE 24 Waiting for news of the Maypole pit disaster in Lancashire, 1908. Note miner's wife in the middle, with headscarf, and ladies in smart hats—probably wives of officials, or the mayoress.

PLATE 25 Mines rescue team, Oakbank, West Lothian, 1926.
Note cumbersome equipment and canary cages.

PLATE 26   The slums built by a coal company.
Back Balby Street, Denaby Main village, near Doncaster. Note row of privies.

PLATE 27   Modern coal-mining at Daw Mill colliery in Warwickshire.

which would have made the owners liable for the quality of the mine discharge into the River Mawddach had the mine remained open.

As ever, the life of the miner and his often numerous family was hard. Wages were around £1 a week and, as we saw with the lead mines of the Pennines, the miners who lived some distance from home spent the week in barracks. Due to the rich Welsh tradition of singing and the influence of Non-conformist religion, their entertainment was largely choral, invoking Welsh songs and splendid hymn tunes. However, the common miner's disease of phthisis, a form of tuberculosis, was exacerbated by the fine dust from drilling in very hard quartz. Life expectancy was probably in the 40s, as it was in most mining districts.

The gold was extracted by various methods of crushing to fine sizes and washing, relying on the high density of gold to deposit its particles on blankets or animal hide. A more efficient method was to amalgamate the gold with mercury, later boiling off the mercury to leave the gold behind; the mercury was condensed and reused. This method is no longer used because of the health hazards associated with mercury.

The Welsh gold mines were Mines Royal, the story of which comes in Chapter 12, which meant that records of output had to be kept with which the payment to the Crown for permission to mine—the 'royalty'—could be calculated. Royalties were paid on 130,000 troy ounces of gold from the 1850s until the mines closed, but it is very likely that much output was never accounted for. In one case, a manager in the late nineteenth century saved £7,000 in 11 months from a salary of £1,500 per year. It was not unknown for a miner to enter the mine by the official entrance but leave by a 'back door' with a nugget of gold in his pocket. Given his wages that must have been a sore temptation.

It is rare for a mine to exhaust all its reserves: there is still gold in Wales, and we'll look at the future possibilities in Chapter 14. However, a nice final note about Welsh gold is that a nugget from Clogau has been used to make wedding rings for Royal marriages from that of the late Queen Mother in 1923 down to the present day. Small amounts of gold are still found in Wales by panning in the rivers. A warning note is that many items sold as 'Welsh gold' frequently contain little, or no, gold from Wales, and a reputable vendor will make that clear.

Britain has no silver mines but has still managed to produce large amounts of the metal. The reason for the anomaly is that galena, the principal ore of lead, usually contains silver: it is called argentiferous galena, and the silver can be recovered by further processing of the lead metal. In fact, some mines have galena so rich in silver that they are called silver mines. The medieval 'silver' mines at Bere Ferrers in Devon were a group of about 12 lead mines, and the lead district of Alston Moor in Cumbria was often called Carlisle's silver mine as the town had a mint authorized by the king to make silver coinage. There were several of these local mints.

British silver was big business, as we can see by looking at the bribes paid by the Saxons to induce the Danes to stop raiding and conquest—the infamous Danegeld, or Danish 'gold'—'geld' is derived from Old Norse 'gjald', or tribute. Payments were made in 991, 994, and 1002; but in 1007 King Aethelred bought off the Danes for two years with 13,400 kg, about 13 tons, of silver. All told about 60 million silver pennies were handed over, and English silver coins of this period are more common in Swedish archaeology than in English. These remarkable figures show not only the sheer wealth of Anglo-Saxon Wessex (which is why the Danes came), but also the efficiency of the Saxon administration.

Argentiferous galena typically contains between 20 and 40 ounces of silver per ton of lead, which may not sound much but, in July 2008, lead was priced at US$1,800 per ton and silver at $17 per ounce. (Metals are usually traded in dollars. Silver is measured in troy ounces, about 2.3 grams.) With as little as 20 ounces per ton, the silver content is worth $340, or nearly 20 per cent of the value of the lead, so extracting the silver is likely to be worthwhile. It was probably even more so when silver was used in coinage: for many centuries, the silver penny was the common medium for paying working people. In 1295, £6,300 worth of silver, in thirteenth-century money, in the form of 1.5 million silver pennies, was transported to Beaumaris to pay the men building the castle; the commander of the escort must have been a worried man. In fact, the output of silver was considerable, with the London Lead Company and the Blackett-Beaumont mines in the Pennines producing 5.5 million ounces of silver between 1725 and 1870.

The trick was to extract the silver without losing the valuable lead and the method used for centuries was cupellation. Ingots of argentiferous lead were heated in a blast of air to about 900°C, and in this oxidizing atmosphere the lead was converted to litharge, lead oxide. Litharge is a red liquid which flowed from the furnace leaving behind a button of fairly pure silver. The litharge can be used to manufacture red lead, a rust-preventing paint, though lead-based paints are now used sparingly because of lead's toxicity. Alternatively, the litharge can be heated with charcoal to remove the oxygen from the lead oxide, a so-called reducing atmosphere, thus recovering the original lead, though some lead is lost to the atmosphere. Cupellation can be used in the laboratory to assay the content of metal samples, but, used on an industrial scale, it is environmentally hazardous, with the fumes causing death to people, animals, and vegetation; moreover the process is wasteful of lead and requires much fuel. Another drawback is that it does not extract all the silver—medieval church roofs have been stolen for the residual silver content of the lead— and is only economic at grades of about 6 ounces per ton.

The growth of scientific metallurgy in the early nineteenth century produced improvements over cupellation. One process was patented in 1833 by Hugh Lees Pattinson; it relies on the fact that the melting point of lead is 327°C and that of silver is nearly 1000°C. Pattinson arranged a set of nine iron pots capable of holding about 10 tons of lead. Argentiferous lead was loaded into the central pot, melted, and then allowed to cool slowly, which caused lead crystals to form on the surface as its temperature dropped below 327°C. This layer of crystals was scooped from the pot into the one on its left while the molten lead in the bottom, which was slightly richer in silver, was poured into the pot on the right. The process was repeated successively, scooping lead crystals to the left and increasingly silver-rich lead to the right. The end result is that the left-hand pot holds practically pure lead and the right-hand pot contains lead very rich in silver which can be recovered by cupellation. While the Pattinson process was based on differential melting points, its practical efficiency depended on the skill and 'eye' of the craftsmen. The combination of science and skill made the process profitable at 2 or 3 ounces per ton. The drawbacks were that it involved a lot of heavy and hot work, with fires constantly being stoked and drawn, using a lot of fuel. It is rather dangerous as well, because being splashed with molten lead is potentially fatal and the men had no protective clothing of the sort foundry workers now wear.

A further, and bigger, improvement occurred in 1850 with the invention by Alexander Parkes of a process that relies on three properties of zinc. One is that liquid zinc does not mix with liquid lead: its density relative to water is 7.5, whereas lead's is 11.3, so the zinc floats on top of the lead. The second is that zinc melts at 419°C and lead at 327°C, so less heat is required than in the cupellation or Pattinson processes. Finally, and most significantly, silver is about 3,000 times more soluble in zinc than it is in lead. Parkes saw that adding about 5 per cent of zinc to molten lead would produce a layer of argentiferous zinc on top of the lead and, when that layer was removed, the zinc could be recovered by, in effect, boiling off the zinc and condensing it in a vacuum, leaving behind practically pure silver.

Rubies, sapphires, emeralds, and diamonds have been sought since the most ancient times, but Britain has none of these. Britain does, though, have three semi-precious stones.

Cairngorm is a variety of quartz, silicon dioxide, which is a fine, smoky-yellow colour and is found in Scotland. It makes very attractive rings and pendants. Rock crystal is a very pure and transparent form of quartz which is used in jewellery, and sometimes for making lenses. A beautiful example is the Alfred Jewel (late ninth century) in the Ashmolean Museum at Oxford. The gold mounting is inscribed in Anglo-Saxon, 'AELFRED MEC HEHT GEWYRCAN', meaning 'Alfred ordered me made'. It was probably used by King Alfred as a pointer when reading the scriptures. Lastly, jet, a very hard and resinous variety of lignite, a low-grade coal, is found near Whitby. Jet can be carved and given a very high polish. Its heyday arrived in 1861 when, after the death of Prince Albert, Queen Victoria went into deep mourning. The diamonds of society ladies, and those who aspired to be thought of as such, were out of fashion, and jet was in as jewellery to be worn in mourning. When the Queen died in 1901 jet's popularity faded, as did much of the employment it had provided. It is increasingly rare, and the genuine article, especially in antique jewellery, is expensive. Beware plastic imitations.

Almost any attractive stone can be cut and polished. Iron pyrite is mainly mined for its iron content, but paler varieties can be used in ornaments and even jewellery. It has been passed off as yellow diamonds, though a gemmologist would instantly see the difference. Malachite is a green-banded ore of copper, and particularly good specimens can be used in costume jewellery and have even been made into tabletops.

11

Accidents and Disasters: The Price Miners Paid

From the earliest mining in the Norfolk flint mines down to very recent times, mining accidents have occurred, and there has been a doleful tale of death and injury, with men, women, and children being killed and maimed. Death and injury were hazards in every type of mining, but coal had the worst record, not only because it was by far the biggest industry but also because of its particular risks. We'll look at coal first and come to metals later in the chapter.

From its inception coal mining had a poor record. One of the dangers is that the roof of a coal seam is usually much weaker than that of a metal mine, and if it is not properly supported, collapses can occur. Explosions are another risk, either when intended blasting goes wrong or when methane is ignited by a spark. Another source of danger is haulage ropes under high tension. At one of the mines I worked in a man was decapitated when a haulage chain snapped and flew down the coal face with vicious force. Underground fires are also dreaded: in 1899 at Whitwick in Leicestershire 35 people, the youngest only 13 years old, died in a fire caused by coal that had been thrown into the gob, or waste because it was too small to be saleable. Breaking into old workings or getting too close to the sea floor can release disastrous floods: in 1815 at Heaton Colliery in Northumberland an inrush of water drowned 75 underground workers. Too often the miners were pressed to produce output by colliery management who were themselves driven by the owners, and the resulting shortcuts, such as inadequate roof support, had inevitable consequences

Mines are very noisy places which can leave one unaware of what is going on. Before fully mechanized roof supports came into service a coal miner relied on the sound of the props creaking (he usually called it 'talking') or the coal 'working' to alert him to a potential roof fall. In a song in a Tyneside dialect known as Pitmatic a miner tells his apprentice to:

> Jowl, jowl, and listen, lad,
> Ye'll hear the coalface workin',
> There's monny a marra missin' lad,
> Because he wadn't listen, lad.

('Jowling' was striking the coal with the handle of the pick, a dull thud indicating a crack; 'monny' means many and a 'marra' was a mate.)

Matters improved somewhat following the establishment of the Inspectorate of Mines in 1850, but, none the less, in 1911 about 2,000 coal miners were killed and many thousands were injured. The case was stated very powerfully by Herbert Smith, president of the Miners Federation of Great Britain, speaking in 1924:

In 1923 ... every working day [on average] more than five persons were killed and 850 men and boys were injured ... Marshall them in one huge procession, four men in a rank, each one yard apart, and you get a procession of injured men stretching a distance of 45 miles. Every 15 yards of that tragic march you would have an ambulance conveying a man who was seriously injured, and every 61 yards a hearse ... This is part of the miners' wages; part of the price he pays in the struggle with natural forces, that the people may have coal, and that he and his family may have bread.

Smith's totals of 1,250 dead and more than 200,000 injured would, indeed, form a column 45 miles long, which would stretch from Glasgow to Edinburgh—about the extent of the Scottish coalfield, but there were some serious accidents in 1923. In one case, seven men were killed and 50 injured when the rope hauling an underground train broke, and at Falkirk 40 men died in an inrush of water. Seven more lost their lives at Nunnery Colliery in Yorkshire in another haulage accident. Again in Scotland, one James Templeton got too close to a coal-cutter and had to have three toes amputated, and a John Adam broke a leg when coal burst from the coalface. Such accidents had dire consequences for the families. For a dead husband a widow might get some compensation, but it was inadequate at best. James Templeton would be lucky to work again, and certainly not on relatively highly paid coalface work. John Adam's income would have vanished for months. In addition to the accidents, there was steady toll on miners' health from lung diseases from dusty conditions, and deterioration of the joints due to the heavy work. The major sources of death and injury were, then, these endless 'minor' cases and ill-health, but some accidents involved so many men that they were truly disasters. The 40 men drowned at Falkirk was one such. However, there were some frightful losses of life due to methane-induced explosions.

Accidents and Disasters: The Price Miners Paid

The special danger in collieries is the methane gas, or 'firedamp', formed by the decay of vegetable matter in ancient swamps. During the formation of the coal, much of the gas is retained within its pores and held there by the pressure of the overlying strata, but mining reduces the pressure on the coal, allowing the methane to escape. As noted earlier, methane burns violently at concentrations above 5 per cent, and, while it does not actually explode, the shock of the flame drives coal dust into the air, and that does explode, causing many serious accidents and disasters. It would be depressing to recount too many stories of these disasters, but I will mention some of them in honour of the dead, and to record the bravery of rescue workers, who sometimes died themselves.

To take a few cases from the Welsh coalfields: in September 1878, 268 men were killed at Abercarn; 176 deaths occurred at Llanerch in 1890; and 276 men and boys died at Cilfynydd in 1894. The collieries at Risca worked the very thick and gassy Black Vein coal seam, and there were 146 deaths in 1860 and another 119 in 1880. The worst disaster ever in British mines happened on 13 October 1913 at Sengenhydd when, despite some 500 men escaping to the surface, no fewer than 436 men and boys were killed, the corpses of about 100 of them having to be abandoned underground. Such disasters were not confined to Wales. In 1880 an explosion at Seaham colliery in Durham claimed 164 lives. In 1908, 76 were lost at Maypole Colliery in Lancashire, in May 1910, 361 were killed at Oaks Colliery in Barnsley, and in December of that year a further 344 died at Hulton in Lancashire. Every coalfield in Britain had its own tragic tales to tell.

The explosion might be initiated by a pick striking metal, an accident with a flame lamp, a spark from electrical machinery, blasting without testing for methane, or (even worse) some fool smoking. The cause of death could be the shock effect of the blast and flying debris, sometimes dismembering the body so that it could not be identified, roof falls following the explosion, scorching from the flame, or slowly choking to death in the carbon dioxide resulting from the blast—the miner's name for this gas is choke-damp. Any ponies underground at the time were also killed, and their carcases would eventually have to be butchered and removed.

When a serious accident occurred the colliery hooter, normally the signal for work to start, would be sounded for several minutes. This tocsin brought the people in a village or small town running to the pithead where they would congregate for hours or even days. Wives and mothers gathered, usually wearing dark headscarves, waiting for news about the fate of husbands or sons, and also wondering how they would survive on what was usually pitifully inadequate compensation. Church organizations such as the local chapel or the Salvation Army often provided bodily sustenance and spiritual comfort during the wait.

It was common for family groups—father, sons, uncles, and cousins—to work as teams, so a serious accident could wipe out all the men in an extended family. To take but one case, the memorial to the Seaham Colliery disaster of 1880, which took 164 lives, has the names of Michael Henderson (Senior), aged 57, and of Roger, aged 29, Michael (Junior), aged 22, and William, who was 19. Until the practice was banned in the 1840s, women and small children worked underground, so the early disasters included them. As to the families, the 1860 catastrophe at Risca left 51 widows and 120 dependent children, and that of 1880 an additional 302 widows and children. An appeal in 1860 raised £7,000 which initially paid £23 per week (no doubt for funerals), although it soon fell to the equivalent of £2.75. By 1880, about £1,000 remained and an appeal for the new disaster also brought in about £7,000. The trustees of these funds now faced the dilemma of caring for the remaining dependants of 1860 or transferring money to the 1880 appeal, and no doubt did their best. However, when the coal mines were nationalized in 1947 it was found that there were many of these funds, the beneficiaries and trustees of which had long since died, and legal work was needed to draw them all into one.

One aspect of disasters was loss of life among the men who went down the pit to attempt rescue in dangerous conditions. The explosion at Cadeby Colliery in South Yorkshire on 8 July 1912 left 35 men missing. Rescue teams, hampered by very cumbersome breathing apparatus and supported by volunteer miners, went down the pit, and a second explosion killed 44 of them. The eventual death toll was 93—some of those brought to the surface subsequently died—leaving 61 widows and 132 children who received no compensation from the Denaby and Cadeby Coal Company, whose policy was to evict a dead miner's family from the company-owned housing within weeks of bereavement. Lacking compensation, the plight of an injured miner or his dependants as late as the nineteenth century could be harsh indeed. Parish relief could be applied for but might be hard to obtain. A local magnate might be generous—as were the Marquess of Anglesey and the Fitzwilliams in Yorkshire—but without such aid the workhouse was the only recourse.

From the mid nineteenth century friendly and provident savings schemes were developed to remedy such uncertainty. Typically a miner paid a few pence each week and the scheme paid a few shillings (usually about 50 pence) per week in benefits for injury, old age, etc. Out of very many, one such was the West Riding Permanent Relief Fund which, by the end of the century, had paid out in excess of £200,000 to some 70,000 widows and children and claimants for infirmity, and, above all, 68,000 accident victims. In due course, commercial firms such as the Prudential offered similar arrangements and the trade unions supported their members in various ways.

Some tragedies were not due to methane but arose from primitive working techniques, as at Hartley Colliery in Northumberland. As was the common practice, the pit was sunk with only one shaft: it saved money. The shaft was divided into halves by a timber partition, the upcast half being used for pumping and the downcast for raising coal and transporting men, by means of a large beam engine. On 16 January 1862, the incoming shift was relieving their predecessors at the coal face, so 204 men and boys were underground and 8 were on their way up the shaft. At 10.30 a.m. the engine beam snapped and 21 tons of cast iron fell into the shaft, doing enormous damage, killing 5 of the men in the shaft and blocking it solid for about 30 feet. Heroic rescue attempts were made, but it took four days to get access to the trapped men. When they were found, their tools showed that they had tried to dig themselves out, but all 204 men and boys had slowly suffocated when the ventilation had ceased. The rescue workers were given medals and payments ranging from £4 to £30 according to the time they had spent in the blocked shaft. A happy, though belated, outcome was the Coal Mines Act of 1872, which stipulated that all mines had to have two shafts giving access to each seam being worked.

This toll of life and damage to collieries could not continue, and in 1911 the government agreed to finance an experimental station in Cumbria to investigate explosions in coal mines. Over the next 20 years that expanded to sites in Buxton and Sheffield for large-scale work, becoming the Safety in Mines Research Establishment (SMRE) in 1947, now part of the Health and Safety Executive's laboratories at Buxton. SMRE designed electrical machines to prevent the sparks inherent in them from passing to a methane-laden atmosphere. New explosives were developed which prevented the flash from the detonation igniting any methane that might be present. More effective and less burdensome breathing apparatus was developed for mine rescue teams, leading to a small self-rescue kit to be carried by every miner, giving him a short-term air supply. There was much additional research into fighting underground fires.

One very significant invention was aimed at suppressing coal-dust explosions. Some wooden planks were suspended from the roofs of the gates leading to the coal face, but were not firmly fastened to the roof. They were then covered with a very thick layer of finely-ground limestone, which was also applied to walls and floors, the theory being that the shock-wave from an explosion would blow the dust into the air and suppress the blast. That was tested in the laboratory and in the experimental mine at SMRE and was proved to work, so it became standard equipment in coal mines. As far as I know it was, fortunately, never called on in practice, as I know from personal experience that the limestone dusting was not always properly maintained. Modern mines use water barriers for the same purpose.

It is, indeed, hard to over-praise this technical work, but it would have had less effect without the framework of legislation, culminating in the Mines and Quarries Act of 1954. That laid down stringent safety requirements, such as the prohibition on taking smoking materials underground. Men are searched before they enter the cage and anyone in violation faces instant dismissal, prosecution, and rather vigorous remonstration from the workmates whose lives he has endangered.

Unhappily, the tale of disasters did not end and, on 29 May 1951, Easington Colliery in Durham suffered an explosion of methane and coal dust caused by sparks from the coal-cutter picks striking pyrite in the coal. Eighty-three lives were lost, of miners ranging in age from 18 to 68. Some of the bodies were unidentifiable and many of the dead are buried in a Garden of Remembrance at Easington Colliery Cemetery.

Perhaps we may leave the last words of this sorry tale of coal's tragedies to a poem, 'The Easington Disaster', written by Fred Ramsey, who was evidently one of the rescue workers.

> At the pithead, in the dawn,
> Rescue teams look old and drawn.
> Sickened by the last vain search,
> Where two men died to save a corpse.
>
> Weeping women, mangled men,
> Can we face it all again?
> Yet, we must, for, if we strive
> Perhaps we'll find just one alive.
>
> Down this grim, blast-shattered mine,
> Bodies sprawl where our lamps shine.
> Once strong men, once eager boys,
> Lie broken like discarded toys.
>
> Our hopes have gone, our tears we've shed,
> There's nothing down here but the dead.
> Eighty three's the final toll;
> The endless, bloody, price of coal.

The power of this lies in the courage of Mr Ramsey and others who went into that mine and had the skill and fortitude to cope with what they found there.

The price of coal was not only paid underground; men and boys were killed and injured on the surface, often in accidents involving the railways that carried coal around the pit and away to market, but one of the most distressing events was the Aberfan disaster of 12 October 1966. It was so horrific that most people who lived through that time can still say what they were doing when the news first broke, much as they can with the assassination of President Kennedy.

Accidents and Disasters: The Price Miners Paid

Merthyr Vale Colliery (a few miles south of Merthyr Tydfil), like all collieries, produced a large amount of useless shale and other waste rock which, after nearly 100 years of mining, was stacked on seven tips. Tip number 7, the disaster tip, had been started in 1958 to store the very fine 'tailings' that remain after coal-cleaning. On the morning of 12 October the tip started to move, and a huge wave of mud slid down Merthyr Mountain. Within a few minutes it wiped out a farm and 20 houses, killing anyone in its path, before, at 9.15 a.m., crashing into Pantglas Junior School and part of the adjacent Senior School. The junior children had just finished singing 'All Things Bright and Beautiful' and had gone to their classrooms, which were on the side of the school exposed to the slip. The slide killed 116 children, mostly aged between 7 and 10, and 5 teachers. Had they been a few minutes later in moving to their classrooms the death toll might have been less. In total 144 people died.

The emergency services responded but were hampered by the sheer scale of the event and by local people digging frantically with their bare hands. That risked making matters worse, and skilled miners from the pit soon arrived, though by then it was too late to save more than a few children. The Chairman of the National Coal Board, Lord Robens, whom we met in Chapter 6, was heavily criticized for not rushing to the scene, but instead going to accept the Chancellorship of the University of Surrey. He offered his resignation, but it was not accepted and he did not insist.

The part of the tip that slid had been sited above a small stream and was saturated with water. Some small disturbance in the tip caused the tailings and water to combine as soft mud—a process called thixotropy—and to flow, triggering the landslide. The men working on the tip saw this and had a telephone, but the wire had been stolen for its copper content. The subsequent inquiry reported that, even had the telephone worked, it would have made no difference to the outcome.

At the inquiry into this tragedy the NCB denied responsibility, claiming that nothing could have been done to prevent the slide. The tribunal disagreed and determined that the *cause* of the slip, as opposed to the triggering event, was 'ignorance, ineptitude, and failure of communication' by the NCB, though none of its employees were dismissed or demoted. The tribunal found that the existence of the stream was known about and that there had been previous minor slips. The Coal Board was ordered to pay £500 compensation for each child, which was then the equivalent of 6 months' wages for a skilled miner.

From there on, tragedy evolved into disgrace. An appeal had raised in excess of £1 million but, after much legal argument, the NCB was allowed to use £150,000 of it to make the tip safe, so avoiding all the costs of doing it themselves. In 1977 the incoming Labour government repaid the money, and in 2007 the Welsh Assembly donated £2 million to the disaster fund. There is a very dignified memorial to the dead. Merthyr Vale colliery closed in 1989.

The price of the riches won from the earth was paid not only by coal miners but also by their fellows in every other sort of mining. In metal mines, the country rock is often very hard and a stope can remain open for centuries, but it still needs watching for loose slabs, and men and boys were killed in roof falls, but also in haulage accidents and in many other ways. Quarries could be dangerous—blasting accidents and rock falls occurred. Large quantities of stone were being moved around, and the place is open to the weather. In the alum quarries on the cliffs near Whitby a boy might be blown into the sea by the North Sea gales, as mentioned earlier.

Sometimes the outcome was less harmful: in 1864 at Killhope lead mine Graham Peart was killed by a rock fall, but his mate, Thomas Rowell, was 'only' trapped. Rowell lived for three days by eating his tallow candles and drinking drops of water. He was rescued and lived to a ripe old age. Overall, and since metal mining was on nothing like the scale of the coal industry and was spared the hazards of methane, the total death and injury tolls were smaller, though none the less disastrous for those concerned and their families.

As with all the mining industries, there was an endless toll of small accidents, but some major disasters also occurred. The Levant mine in Cornwall was typical of many in that the miners had to use ladders to get to and from the shaft bottom as well as walking a considerable distance underground, on top of walking to the mine in the first place. In a mine 1,000 feet deep—and many were much deeper—climbing the ladders at the end of the shift was equivalent to scaling a very tall television mast. Hardly surprising, then, that the working life of a miner was short, and men must sometimes have fallen from the ladders through sheer exhaustion. A great advance at Levant was the construction in 1857 of a man-engine in the shaft. As we saw in Chapter 4, this was a square wooden rod, made up in sections to a total depth of 1,200 feet. At 12-foot intervals a stoller, or step, was fitted to the rods, with corresponding steps on the side of the shaft. As the beam engine on the surface raised and lowered the rod the men stepped from shaft to rod, and back again, and so to the surface.

The system worked well for 62 years until, in 1919, a steel cap on the beam at the surface snapped and 31 men and boys plunged to their deaths. Retrieving the bodies and clearing the heavy rods from the shaft was a monumental task, but recovery of his mates is always the miner's first concern. One man was rescued after two days, but sadly died at the surface. How well the man-rider had been maintained is an open question. Man-carrying cages had been in use elsewhere for years, but it is too simple to say that cages should have been in use at Levant by 1919, as, apart from the cost and loss of production—Levant was a very rich mine—there is an innate conservatism in many that resists replacing a system that seems to work.

After the accident the mine continued until 1931, but working from the shallower levels.

The very wet mines of the Pennines and Cornwall are free of methane but had the hazard of drowning, especially where the workings extended under the sea, as was common in parts of Cornwall. In one case, seven men were killed in August 1858, at the Porkellis United mine in Cornwall, when a surface pond that stored the very fine slimes that remain after ore has been crushed, ground, and processed, collapsed and flooded the mine. A disaster even worse than Levant happened in 1846 at East Wheal Rose, also in Cornwall, when an inrush of water flooded the mine and drowned 36 men and boys. Water was not the only hazard, and in 1900 an underground fire at Snaefell lead mine on the Isle of Man, caused by a naked light igniting timber, claimed 20 lives.

Modern British mines are as safe as they possibly can be, and, while accidents still happen, the record is one of the best in the world, and is infinitely better per day worked or per ton produced than, say, the record of China's coal mines. It is impossible to conceive of another Sengenhydd or Levant, and the old coal tips have been cleared away or stabilized so we need not fear another Aberfan. However, next time you buy some coal or an antique copper kettle, perhaps you will spare a thought for what it really cost.

12

Lords and Adventurers: The Mine Owners

There have always been people who owned mines and developed new ones. In Neolithic, Bronze, and Iron Age times it might well have been a tribal leader or local 'king'. The Roman authorities certainly controlled gold, coal, iron, and lead mines, and the later Saxon kings exerted some form of control, without which they would not have been able to pay the Danegeld. The first real, national, ownership came with the Norman conquest (which we'll look at in a moment) but by the eighteenth and nineteenth centuries many of the owners of mineral rights had become rich, and some established dynasties from father to son through generations—though the succession might fail through the lack of a male heir or because of his incompetence or lack of interest (women having, with few exceptions, no role in the business). It is hard today to appreciate the power and status that these people had. Until the huge social changes after the First World War, men doffed their caps and women attempted a curtsey when the local magnate or his family passed by.

The owners were usually referred to as mineral lords—some also had peerages—and held the land under which the mining took place.

In some cases, the land had for generations been rented to tenant farmers, who were, at least in principle, seen by the owner as his people, to whom he had a duty of care. On the other hand the adventurers—it is a Cornish term meaning shareholder—simply leased land from the owner, operated the mines, and paid him a royalty on tonnage mined.

Some owners were also adventurers, and a successful adventurer might rise to ownership. If the owner operated mines on his own land, via his agents and managers, the personal bond between the owner and the agricultural workers might be carried over to the new miners and their families, as we shall see with the Fitzwilliams. Some of the owners and adventurers were also in the forefront of technology, developing new methods of mining and refining, or even whole new industries, such as shipping and ports, because it was in their interest to do so. For instance, the Marquess of Bute was immensely rich from royalties on his lands in the Glamorganshire coalfields, and he used his wealth to develop Cardiff docks to handle coal exports. He also restored Cardiff Castle and endowed the city in other ways.

The power of the mine owners lay in their control of employment, as a dismissed miner not only lost his job but he would be likely also to lose his home if he lived in a company cottage. In such a case he would have little chance of finding work locally, and either had to walk to another mining area in the hope of getting a job, or be dependent on parish relief or charity. Many miners had large families—8 or 10 surviving children was not uncommon—so unemployment led to great hardship for the whole family.

The role of the Crown as an owner started in 1066 when William I, the Conqueror, declared that he owned *all* land and minerals. The point of the Domesday Book in 1086 was to work out what he did own, the mines becoming known as Mines Royal. Over the next centuries he and his successors rewarded powerful subjects with grants of land, though not always with the mineral rights, and some of those grants of land and/or rights were sold, over time, to others. In general, anyone who mined anything had to pay a royalty to the king until that was abolished in 1689 as a revolt against the hated tax of the Stuart kings. However, all gold and silver mines were retained by the Crown and are still known as Mines Royal.

Nowadays, Mines Royal are part of the Crown Estate which comprises about 250,000 acres of agricultural land, property in London and elsewhere, and another million acres where the Crown owns the mineral rights but not the surface. The Estate includes about 1,000 miles of the shore between high water and the 12-mile territorial limit, so that mining under the sea, as at Boulby Potash, is of financial interest to the Estate. The Crown also has the right to the mineral resources of the UK continental shelf. Royalties have to be paid to the Estate for any mineral mined on its property. However, a licence is required for exploration for gold and silver anywhere in Britain and an annual rental plus a royalty on the output have to be paid. A licence is also required for leisure panning for gold, as well as permission from the landowner. The administration of all this complexity is the role of the Crown Estate Mineral Agent. Given the connection between the Royal School of Mines and Prince Albert, the Agent was usually an RSM engineer, but he is now a partner in a firm of consulting mining engineers.

The considerable revenues from the Estate's agriculture, property, and mining go to the Exchequer in exchange for the Civil List that pays the monarch's expenses and supports senior members of the Royal family, with the Exchequer getting much the better part of the deal. Perhaps oddly, the Crown Estate belongs neither to the government nor to the monarch in person, but is part of the hereditary possessions of the sovereign 'in right of the Crown'. The complicated, but effective, system of the Crown Estate has evolved over the past 1,000 years from the Norman conquest to the present day in the usual pragmatic, commonsense English way. Scotland's Mines Royal are also administered by the Crown Estate Mineral Agent, but within the framework of Scots law.

Some of the private mine owners inherited riches, while others had to create their own wealth. We shall look at a few of them, starting with the Earls Fitzwilliam, who held land near Rotherham in South Yorkshire for generations. By the 1720s their wealth had enabled them to build Wentworth Woodhouse Hall, which was the largest family house in England. The facade is twice as wide as that of Buckingham Palace and the roof is four acres, contributing to the demand for lead that we saw in Chapter 5. The surrounding land was on top of the very rich Barnsley Main coal seam where collieries had opened by 1750. These were inherited by the fourth Earl, adding to the family fortune.

The sixth Earl (1813–1902), Thomas, had eight sons and six daughters, all but one son surviving infancy. Unhappily for the Fitzwilliams most of these produced no children, and, more seriously for the succession, none produced a son. Thomas was slightly eccentric: even in 1902 there was only one water closet in the house, no central heating, and no electricity. His death was a major occasion in Wentworth and the surrounding area, with many hundreds of people lining the route. He had lain in state in the house; whereas the dead in a pit village were often stood in a corner until the funeral, or even shared a bed with the living.

Thomas's son had predeceased him, so he was succeeded by his grandson, William Fitzwilliam, who was called 'Billy Fitzbilly' by the miners. As well as owning Elsecar and New Stubbin collieries, he also had interests in chemicals and the Sheffield Simplex motor car, intended to rival the Rolls-Royce. William became a qualified mining engineer, took an active interest in mine safety, and provided pensions and injury compensation for his workers. The Fitzwilliam villages had gardens and vegetable plots unlike those of the neighbouring Denaby and Cadeby Coal Company, whose villages were poorly built and squalid, and who evicted the families of dead miners, and to whom compensation and pensions were unknown. Fitzwilliam, by contrast, provided meals for the miners' children during the great depression of the 1930s, and his Countess, Lady Maud, worked alongside the miners' wives to serve the food. After his death the title passed to a son and two cousins, and the earldom died out in 1979.

During the Second World War, the estate was progressively destroyed by open-cast coal mining to meet the desperate need for fuel, but the final indignity occurred in 1946 when the Minister for Fuel in the Labour government that nationalized the collieries, Emmanuel 'Manny' Shinwell, decided to mine coal right up to the door of the house, with the spoil piled 50 feet high up to the eighth Earl's bedroom window. Mining engineers and geologists from Sheffield University reported that the surface coal was of very poor quality, which was true, and that drift mining by tunnels from the surface was a better option, and the Yorkshire Miners' Union objected because of their appreciation of the Fitzwilliams as excellent employers, but to little effect. A sad end to a dynasty who seem to have done much good, and who appear to have been genuinely loved and respected by their thousands of employees on farms, in factories, and 'down t'pit'.

Thomas Williams (1737–1802) was a successful lawyer in Anglesey, becoming what we would now call the family solicitor to the notables of the island. In later years he became known as the Copper King because he eventually controlled copper mining in Anglesey and, through an agreement with the Cornish mines, personally decided how much metal was to be produced and the price at which it was sold; these powers led to a parliamentary enquiry.

His breakthrough to fame and wealth came in 1761 when abundant copper was discovered on Parys mountain. The mountain was poor agricultural land, so it had not really mattered who owned which parts of it, but it mattered a very great deal when parts of the mountain, of which Sir Nicholas Bayly asserted ownership, proved to be rich in copper. Bayly's claim was challenged by another family and the matter was fought in the courts for a total of seven years. During this period mining continued, but mainly picking out the richer ore, leaving the mines in poor condition. More effective management was needed, and the outcome of the lawsuit was that Williams controlled the Parys mine and, later, the Mona mine which had opened on the eastern part of the mountain.

A significant use of copper around 1760–1800 was, as we have already seen, to sheath the Royal Navy's ships against teredo worm and clinging weed—very important, since Britain was at war with both France and the American revolutionaries. However, the copper sheets were fastened with iron nails which effectively dissolved due to the slight electric current created in sea water between copper and iron. This became so much of a problem that the navy was on the point of abandoning sheathing, which would have done no good to Williams's expanding business. Fortunately, a method of making copper nails and bolts was developed, which saved both the navy's bacon and a large part of his. Williams became involved in making these items, and he was not averse to selling the bolts to both France and Spain, until at the end of the eighteenth century Britain was actually at war with them. By then the copper bolts had been used to build their ships.

The copper industry in the late eighteenth century was little short of chaotic. Apart from Anglesey there were numerous mines in Cornwall and Devon, and lesser producers elsewhere, all of whom had to deal with a relatively small number of smelters in Cornwall and, mainly, Swansea, the latter having the big advantage of access to cheap and plentiful coal. Williams's talent for business rose to this and, after incessant travelling between London, Cornwall, Birmingham, and Anglesey, in 1785 he brought about what became known as the Great Treaty. This divided the markets at Liverpool, Bristol, Birmingham, and London between Cornwall and Anglesey, including an agreement with the smelters. The aim was increased prices, which was not popular with industrialists such as Birmingham's Matthew Boulton, who manufactured all manner of copper items such as buttons and coins. The advantage to Williams was that he could achieve both higher prices for Anglesey copper and lower production costs. When Williams took control of the Cornish trade the county's annual output was restricted to 3,000 tons, which caused mines to close and, as we saw in Chapter 3, led to ferocious riots.

The Williams monopoly lasted from 1788 to 1792, during which time Anglesey's output was falling, leading to his placing more emphasis on manufacturing. This generated further conflict with Boulton. At the height of his power, Williams controlled the two Anglesey mines, and owned other businesses, such as smelting in South Wales and Lancashire, chemicals in Liverpool, copper- and brass-making in Flintshire and the Thames Valley (where he built a mansion at Great Marlow), plus three banks. All this represented a capital of about £800,000. Williams, who had always been asthmatic, in 1802 contracted gout, eventually dying on 27 November 1802. His sons, Owen and John, continued with mining until 1810 and finally left the copper business in 1825. Owen's six granddaughters all married into titled families, largely due to the wealth accumulated by Thomas Williams, who had been a truly remarkable man.

The Greaves family of the slate industry of North Wales were a dynasty of adventurers who rose to great wealth but, unlike some of the coal companies, treated their workers as human beings. The family originated in Warwickshire and were successful bankers. However, John Whitehead Greaves (1807–80) was advised to seek his fortune in Canada, and at the age of 23 set out for Carnarvon to get a cheap Atlantic passage. At Carnarvon he was so impressed by the commercial bustle of the slate trade that he abandoned ideas of Canada, and by 1833 went into partnership with a man called Shelton and took out a lease to explore the Upper Glynrhonwy quarry near Llanberis in the slate-mining and quarrying area of Blaenau Ffestiniog—despite the fact that Glynrhonwy had bankrupted the existing lease-holder. However, Greaves suspected that Bowydd quarry, which was included in the Glynrhonwy lease, was on the verge of finding good slate, and this proved to be the case. The next few years were good for Greaves: the Ffestiniog railway was built to transport slate to Porthmadoc, banks opened in response to the area's new wealth, and the destruction of much of Hamburg in 1842 by fire created a huge demand for Welsh slate.

Greaves thought that very high quality slate lay under the neighbouring Llechwedd property, but it took three years to negotiate the lease from the people who had grazing rights on the land. Exploration began but, by 1849, Greaves was in serious financial difficulties and could no longer pay his workforce. Two families of miners offered to continue to work for nothing as long as the stock of gunpowder lasted, and a few weeks later discovered the Old Merioneth Vein of excellent slate. The mining families were rewarded with free housing and guaranteed employment. The promise was honoured as late as 1985, when a descendant turned up and asked for work.

In 1850 Greaves invented a sawing machine to cut slate into manageable blocks, using that queen of power sources, the waterwheel. He was now well established. Llechwedd slate won prizes at the Great

Exhibition of 1851 and was sold worldwide. In the following years the Greaves family built 12 ships to carry their products to market. Another step forward came in 1853 when, after much negotiation, an incline railway was built to connect Llechwedd to the Ffestiniog railway, making redundant the slow and costly packhorses used until then.

Greaves retired in 1870 and his son, John Ernest, aged 23, took over the firm together with his brother Richard, who had trained as a mechanical engineer. Richard invented a slate-dressing machine which is now in worldwide use. By 1900 the Greaves' business employed 600 men, and the family were rich, but hard times arrived when a protracted lockout at the Penrhyn quarries and aggressive marketing of man-made roofing tiles depressed the slate industry, to the point where Llechwedd employed only 150. The board of directors recognized that increased mechanization was needed and recruited Martyn Williams-Ellis, a qualified mechanical and electrical engineer and nephew of John Ernest and Richard, though it was not until the end of the First World War that he took up his post. Williams-Ellis introduced hydro electric power, using Llechwedd's own dams, and in 1931 persuaded the board to switch to open-cast quarrying instead of underground mining. Llechwedd was back on a sound footing, and it still produces all manner of slate items, but in 1972 it also became a significant tourist attraction. All told, a very successful dynasty of mine adventurers.

As we saw in Chapter 8, a salt industry has existed in and around Northwich since at least Roman times, but it was not until the mid eighteenth century that it started to evolve into the large-scale and organized industry that still flourishes. Throughout much of its history there seem to have been many salt owners, mostly on a small scale, and the Marshall family were unusual in maintaining a family succession for five generations.

The first Marshall, Thomas, moved from Nantwich to Northwich in about 1720. He served an apprenticeship in this profitable industry and in 1734 started out on his own account. He must have been a man of determination and commercial acumen as he soon acquired property in Northwich for offices, and facilities for boiling brine. In a further expansion he leased brine pits and bought land under which there were deposits of rock salt. Not content with that, he married an heiress, and, after 40 years in the salt business, the Marshall family fortune was well established.

Marshall's son, another Thomas, lived from 1735 to 1797 and was as ambitious, astute, and determined as his father. In 1781 another bed of rock salt was discovered and he bought land on which the famous Dunkirk mine was developed, and which exported salt to Russia and elsewhere. At this time, salt was being produced by boiling brine using coal as a fuel, and Thomas II took the bold step of leasing collieries in Lancashire, improving the canals, and transporting coal on his own barges, thus securing supply and controlling costs. He also introduced the new Boulton and Watt steam engine to raise salt from the deeper levels of the rock salt mines. Perhaps his most significant contribution was to bring about at least a degree of collaboration between the salt firms in order to reduce over-production. He lived in interesting and turbulent times, dying only four years after the execution of Louis XVI in the French Revolution, but by then the Marshalls were very rich. Evidence of their wealth survives in Northwich in the shape of two mansions, an elegant townhouse, and a parish church.

Thomas II was succeeded by his sons, John (1765–1833) and Thomas III (1767–1831), who built up the business to become the biggest salt-trading company in Cheshire. The firm survived a recession and bought more land, built workers' housing, and acquired a shipyard. John and Thomas III continued their father's work and were significant figures in the formation of the Salt Trader's Association in 1805—the year of Trafalgar—to fix prices and production quotas, something that nowadays would be illegal. Up to this point the Marshalls had survived into their sixties, quite long by the standards of the age, and almost certainly longer than most of the salt workers, but the fourth Thomas (1802–38) lived for only seven more years after he took over the business in 1831. He seems to have had interests outside of salt, and the firm was run by his mother. His son, Thomas Horatio, was only 5 years old when his father died, so the business was run by his guardians, who appear to have been somewhat uninterested and the Marshall influence declined. By the time Thomas Horatio came of age he had decided on an army career, and the mines and works were sold. He was able to leave considerable wealth to his children, but the value of that fell sharply after 1914, a slightly sad end to this saga of endeavour, hard work, and foresight.

There were many other families in coal, lead, iron, and everything else, that we have no space to describe. A few of them were the Vivians of Cornwall, the Blackett-Beaumonts in the Pennine lead fields, the Briggs in Yorkshire coal, the Wynns of Gwydr in Welsh lead-mining, and the Robinsons in Cumbrian gypsum.

The mine owners, of whom we have mentioned only a few, were also employers, and before we look in the next chapter at the lives of the miners, we need to say something about the relationships between what the Victorians would have called 'master and man'. That would require a book of its own—several are listed in the bibliography—so we can only pick out some main themes.

One is that, as we have mentioned, monasteries owned many mines. Monks held services, copied manuscripts, gave alms, and cared for the sick, while lay brothers worked the land and the mines. Here there was no master and man, and lay bothers were guaranteed care for life. Being one was seen as a good lifestyle choice, despite celibacy and the requirement to attend some services, but a brother who erred would have to undergo some penance.

In the 1530s, the monasteries were abolished and more conventional owners took charge. During the next few centuries thousands of them came and went, and it is hard to generalize about their relationships with, and behaviour towards, their workers, but the picture becomes clearer by the mid nineteenth century when the mining industries were large and mature. The rich landowners often dealt with their agricultural and mining employees via an agent, or 'man of business', and had little or no direct contact with the workers. Two exceptions were William Fitzwilliam who, as mentioned above, qualified as a mining engineer, and, in a different way, Lord Penrhyn in the slate industry, who fought bitterly to break the quarrymen's union as it smacked of radicalism, though he provided a hospital and some pensions. The owners of mining companies were often tyrannical, but as trade unions developed they were willing to deal with them rather than face the chaos of an uncontrolled strike.

The more responsible owners, such as the London Lead Company, which had Quaker origins, made good provision for their workers at places such as Middleton-in-Teesdale. In 1824 cottages were built, each with its own garden. This saved the more deserving people from the rising rents caused by the increased population of the area as the lead industry expanded. The tenants were required to abstain from intoxicating drink. Other owners, whether landowners or mining companies, supported brass bands and schools, and built chapels as part of their housing schemes. Many of the bands, particularly in the larger collieries, which could draw on a large pool of potential musicians, were very good indeed. Some bands still exist, even though their colliery has closed. Regrettably, the worst of the owners did nothing. As we have seen, some companies evicted the family of a dead miner.

Lords and Adventurers: The Mine Owners

A common feature of all these nineteenth-century owners, even well into the twentieth, was an almost complete lack of social intercourse apart, perhaps, from an occasional appearance at Christmas. Housing was strictly segregated, so that a mine manager would get a double-fronted house, with spacious rooms and accommodation for a servant or two. As we shall see in the next chapter, a coal miner's family were less amply catered for.

One factor in social relations was the influence of religion, which, even well into the twentieth century, had a power that nowadays we find hard to understand. It was normal for even a small town of 10,000 people to have six or eight dissenting chapels of various persuasions. A second Anglican parish church might be added as the town grew, and there might also be a Catholic church, depending on the number of people of Irish origin. These places of worship were, however informally, themselves segregated into what we would now call up- and downmarket congregations. Until late in the nineteenth century prosperous people rented a pew, reserved for their exclusive use, and as near to the front as possible.

The whole system was encapsulated in a verse of the children's hymn, 'All Things Bright and Beautiful', written in 1848 by Mrs Cecil Frances Alexander, the wife of an Anglican clergyman who later became a bishop.

> The rich man in his castle,
> The poor man at his gate,
> GOD made them, high or lowly,
> And ordered their estate.

The verse is no longer included in hymn books, but it certainly taught the children who sang it where they stood in life's pecking order.

13

The Mining People and Their Communities

For more than 4,000 years Britain's miners have, to quote Job, with whom we started this book, brought to the light fabulous mineral riches. But to create that wealth the miner toiled back-breakingly hard, was often harshly treated by unscrupulous managers and owners, and he and his family frequently endured great hardship and poverty. It is a tribute to the human spirit that, despite their tribulations, these extraordinary people did not succumb to brutishness. While there were drunkards, thugs, thieves, and even murderers, there were also those who strove for education against all the odds, and even more so for their children. There were fine musicians, eloquent preachers, and outstanding individuals who stood up for the rights and welfare of others. For instance, Ashington in Northumberland, which called itself the biggest mining village in the world, boasted a philosophical society with regular lectures, operatic and dramatic societies, and much else, including a miners' painting society, the works of which are now exhibited at Woodhorn Colliery Museum. In short, the mining people managed to remain human, in spite of terrible adversity, and that is perhaps their greatest achievement. In this chapter we will look at the lives these people led, though perhaps one needs to have worked in their communities to fully understand mining people.

I have said several times that the life of the miner was all too often short: death by the age of 45 years was common. Men and women doing such grinding labour, often in the open in all weathers, could scarcely hope to see three score years and ten, and data from the Registrar General's Annual Reviews for 1851 and 1861, supplied by the Office for National Statistics, tell a doleful tale. Table 13.1 picks out some categories of people and shows, by calculating from the Registrar's meticulously compiled data, the percentage of males in that category who have died by a given age. The figures are approximate but they show the comparisons.

'Persons of rank or property' were the landed gentry; the second category is farm owners or tenants; labourers include farm and unskilled

TABLE 13.1 Deaths by social category, 1851 and 1861.

CATEGORY OF PERSON	PERCENTAGE DIED BY AGE CATEGORY						
	20	25	35	45	55	65	75 AND OVER
Persons of rank or property not returned under any office or occupation	2	7	13	24	42	70	100
Persons possessing or working the land and engaged in growing grain, fruits, animals, and other products.	6	14	22	32	49	74	100
Labourer (branch of labour undefined)	9	12	35	50	71	87	100
Miner (mineral not stated)	18	41	59	73	92	99	100

factory workers. The deaths of miners rarely stated the branch of mining, and those that do are too few for reliable calculation, so we have a catch-all category. The causes of death would be everything from infant mortality, malnutrition, accident, and disease, to sheer exhaustion, but the picture is clear: the miner had almost no chance of living beyond the age of 55, while his more fortunate brethren, or those in less arduous or dangerous work, lived a good deal longer.

Statistics show outcomes, but to get a glimpse into daily life we need a written record. However, the vast majority of miners were illiterate until the expansion of education in the nineteenth century. Even then, boys left school as soon as possible with the minimum of learning, and in any case at the end of a day of hard labour literary endeavours cannot have been a high priority. It would take a truly remarkable man to write a miner's autobiography; one such was Bert Coombes, a Welsh collier who wrote *These Poor Hands*, first published in 1939.

Coombes was born in 1893 in Glamorgan and brought up in Herefordshire. After he left school he worked on the family's small farm from 5.00 a.m. to 10.00 p.m. His father had the contract to cart coal for a local charity, and was paid with a ton of coal at Christmas, which was about the only time of year that the family could be warm and dry. In a sense they were fortunate, as Coombes was the only child; a neighbour had to keep a family of nine on 16 shillings (80p) a week. From the farm the fires of the furnaces at Ebbw Vale steel-works could just be seen, so, in search of warmth, at the age of 18 Coombes set out by train to find work in the mines, just as a friend of his had done. He spent most of the rest of his life in Resolfen, Glamorgan, though he gives the village a fictional name in the book. His first shock was to find that swedes were sold in Resolfen for a penny (less than 0.25p) a pound, which was 15 times what he and his father had been paid for not only harvesting the vegetables but also loading them into a cart.

Coombes got a job as a miner's helper by falsely claiming that he had previously worked in collieries, and so began a lifetime of underground work, at one point in a seam only 15 inches thick, where he lay on his belly and moved in a kind of swimming motion. His descriptions of a miner's life are accurate and compelling: appalling conditions, inadequate ventilation, and terrible accidents, all on top of walking three miles up a mountain to get to work, even in winter. He also writes of the companionship of miners with the occasional humorous episode. For instance, cloth caps were all that protected the head, but in the 1920s ex-army steel helmets became available and, to demonstrate their effectiveness, the deputy in charge of the coal face was banging a volunteer on the head with a lump of wood when the victim, thinking the demonstration was over, took off the helmet and was promptly laid out by another clunk.

Coombes was an intelligent man who rapidly mastered the complexity of the first coal-cutter installed at his pit and so became a machine man. This later stood him in good stead when work was short for ordinary colliers. During the general strike of 1926 he made a violin, learned to play on the basis of a promise to pay for the lessons when he was next in work, and became an accomplished musician, founded a cricket club, and qualified as a first-aid man, though he had to pay for lessons and for travel. Quite what Mrs Coombes thought of these additions to a hard-pressed family budget he does not reveal.

The real power of Coombes' autobiography lies in his descriptions of how the miners were treated by their employers. The men were paid by the piece: so much per ton of coal, per prop erected, and so on; at the end of the week, measurers came round to calculate what was due, but with inevitable attempts to avoid payment. However, anyone making too much of a fuss or perceived as insolent might be dismissed on the spot, losing his company housing at the same time. The miners did not trust the owners, so the local branch of the Miners Federation elected a 'check weighman' to verify the company's tally of output; Coombes' father-in-law was one such. Coombes' colliery came out on strike for three months in support of a dispute over the replacement of a check weighman at another colliery. When they went back, his father-in-law was a marked man for his activities during the strike, and did not work for another five years.

A major dispute occurred early in 1921 when the owners, having stocked coal, locked the men out. The men had no money and depended on parish relief, and Coombes writes bitterly about the arrogance and harshness of the relief officers. To save some money, he soled his boots with old car tyres. After 9 months the dispute ended with the men returning to work with longer hours and less pay. But the really big catastrophe for the miners was the general strike of 1926. Coombes helped to make ends meet by using his first-aid training; he and his mates would sometimes fake a car accident and collect donations for the 'injured' from passing motorists. The men were offered work to keep the pit in good order but they would not break the strike. The owners brought in what the strikers called 'blackleg' labour with a heavy police escort, paying them 15 shillings (75p) a day when they wanted to cut the regular wage to 10 shillings. The blacklegs blatantly stole the miners' tools, but any protest invited arrest. A set of tools cost at least £4, a substantial sum.

It was, however, during these years that Coombes started to write short stories and articles for left-wing and trade-union publications. Success came with *These Poor Hands*, which sold 80,000 copies within a year of publication, and the income from writing enabled him to start a small farm.

The Mining People and Their Communities

Mr Coombes died in 1974 and we should let him have the last word about a miner's life in the 1920s. 'I feel that I hate the continual slavery and the dust; the poor clothes and bare living; the need for decent men to beg their bread [when there was no work].'

Away from work, miners and their families often lived in dreadful conditions, even as late as the 1930s. As we saw in Chapter 8, some of the houses in the salt-producing area of Cheshire were built of the ash from the salt boiling. Miners in the lead districts of Yorkshire and Derbyshire had cottages of local stone; basic but slightly more tolerable, as many had a plot of land as a small holding. The most widespread distress, though, was felt in the coal-mining areas, not least because there were just so many of them: a million men worked in coal-mining in 1900, and there were many hundreds of collieries and uncounted scores of mining villages. To be fair, some of these had cottages of reasonable quality, with vegetable plots and gardens, but the worst can only be described as hell-holes, it being in the colliery owners' interests to house as many people as possible, at the least cost, on the smallest area of land.

The result, all too often, was squalid rows of back-to-back terraces, some of which still existed as late as the 1950s. In a back-to-back terrace the front half of the building is one house, and there is a separating wall, on the other side of which is another house with an access door from the next street. The usual layout was a downstairs room and one bedroom—'one up and one down', as it was called. Sometimes it was on three floors, or 'two up and one down'. One side of the house was designated to be the 'front', and in alternate rows the fronts faced each other across a street, which might or might not be paved. The backs also faced each other across a back lane, down which ran a row of earth privies which were emptied about once a week by 'night soil men'. The consequence was that anyone living in the front part of a house had to walk to the end of the street and then to their privy which, if they lived in the middle of a row, could be 200 yards. Once there they might have to wait in a queue, as privies could be shared by as many as 30 people. Inevitably, many houses had stinking chamberpots waiting to be emptied, and not always in the proper place.

The houses were rented from the colliery company or from private landlords and were usually in poor condition. The slum clearance programme in the 1930s often made matters worse. The town council would designate property as unfit for human habitation, but during the economic depression of the 1930s they had no money to build what we would now call council houses. In that case, the landlord would be even less willing to pay for repairs, and the poor family was stuck with the result. We should say families (plural), as it was not unknown for two groups to share one side of a back-to-back. Every drop of water had to be carried from a tap at the end of the street.

In 1936 George Orwell was commissioned to spend two months living in Wigan and to write about his experiences. The first half of the resulting *Road to Wigan Pier* is a devastating account of what he saw. One of its most moving passages concerns a woman that Orwell saw who was trying to clear a blocked drain. She was probably 25 but looked 40, and he goes on to write, 'For what I saw in her face was not the ignorant suffering of an animal. She knew well enough what was happening to her—understood as well as I did how dreadful a destiny it was to be kneeling there in the bitter cold on the slimy stones of a slum back yard, poking a stick up a foul drainpipe' (Orwell, p. 15).

In the latter part of the nineteenth century the normal terraced house came into use, and many thousands still exist. Typically it had a front door, a front room, a door to a back room, and a back door—

'two (or three) up and two down'. Piped water was provided, together with gas light. There was usually a paved back yard, an outhouse for coal and another as an outdoor flushing toilet, though bathing was still in a tin tub in front of the fire. Each terrace had a corner shop and some houses had a slot in the outside wall for the boards on which the dead were dealt with by the 'laying out woman'. Laundry could be a problem as clothes could only be hung out to dry when the wind was not blowing coal dust from the direction of the colliery. Nowadays these houses have been thoroughly modernized and can command high prices.

The wife or mother of a miner had to feed him and their often numerous family, usually on a very restricted budget. Old women in the mining districts used to talk of buying 'a pennorth of pot stuff', meaning that for a penny (about 0.4p) she could buy a collection of vegetables, and possibly some scraps of meat, sufficient to make a stew of some sort. Modern nutritional standards suggest that some-one doing heavy manual work needs about 3,500 calories per day, but it is quite clear that, given the money available, most mine workers got nothing like that amount. Perhaps fortunately for the women, appetites were often not large after several hours of working in very bad ventilation.

Nowadays, some of the culinary creations of the housewife have become standard fare. The famous Yorkshire pudding is, ideally, the size of a plate, and was eaten as a course in its own right, to fill the miner up before a scanty meat course. The 'pot stuff' would make a version of 'Lancashire hot-pot'. Tripe and onions were cheap, and other favourites were haggis and black pudding. Cornish pasties could be taken down the mine, sometimes made with meat at one end and a sweet at the other. The pastry crimping enabled it to be eaten with dirty hands.

All of these can fairly be called 'poor man's food'—pizza, Dutch pancakes, and quiche Lorraine, named for a mining province in France, are others, many of which are now seen as delicacies. Bakeries in Cornwall vie for the prize for the best pasty, as they do in Yorkshire and Leicestershire for their pork pies. A top hotel in Edinburgh offers excellent haggis, 'tatties and neeps' (potatoes and turnips), and an expensive, upmarket restaurant in London, which specializes in traditional British food, serves outstanding Yorkshire puddings—a far cry from the hard-pressed wife trying to feed a family on next to no money.

The miner's recreation was limited by poverty and exhaustion until conditions and wages improved after the Second World War. Until then, as we shall see below, the Methodist and Baptist chapels played a big role, with various celebrations, and an emphasis on temperance. Sport was popular, with the League version of rugby being strong in the northern mining districts. There is an element of truth in the cloth-capped whippet fancier, and pigeon racing attracted fanatical support. Naturally, the pub, or the Miner's Welfare, where beer was cheaper, featured strongly, especially where men worked in a hot pit. It was quite common for women to go the pit on pay-day and demand the pay packet lest her man drink it all and leave her with nothing except the pawn shop. Recreation for miners' wives was almost non-existent, apart from the chapel functions or, in later years, the bingo hall and television. As we have seen, it was common—almost a right—for son to follow father into the mine or quarry, so many families lived in the same locality for generations. That provided an extended family of relatives within a short distance, giving opportunities for childcare when the mother was at work in, say, a Lancashire cotton mill. The close confines of the mining villages also provided neighbours, so the miner's wife was not deprived of company and support in illness. Unfortunately, these closely confined relationships could also spark bitter quarrels and long-lasting feuds.

The lives of mining people country-wide were affected when, from the mid eighteenth century, Britain experienced a religious revival inspired to a large extent by John and Charles Wesley. John's evangelism was so enthusiastic and vigorous that the Anglican hierarchy barred him from using its churches. That led him to preach throughout the country, often to an initially rowdy reception, and the Methodist Church came into being, independent of Anglicanism. Methodism became hugely influential in industrial and mining districts, and was very strong in Cornwall, though in much of Wales the Baptist Church was favoured, and Scotland had long had its own dissenting churches.

One reason for Methodism's impact on these impoverished and suffering people is that there are many stirring hymns, the words of which perfectly caught the plight of the miner, the steel- and mill-workers, fishermen, and many others. Take, for instance, 'What a *friend* we have in Jesus', or 'I will cling to the old rugged Cross, 'til *my troubles at last I lay down*', with their combination of the promise of spiritual salvation and comfort in life's tribulations, and it is not surprising that Methodism became a huge success.

Methodist and Baptist chapels are a striking part of the mining landscape: even small villages had a chapel. Some were small and plain, others very beautiful and elegant. The buildings were paid for by the mining company or by a local philanthropist, or were built by the people themselves. When new mining villages or industrial districts were built, it was normal for the builder to include a chapel. Apart from their religious function most had a Sunday school, which was important before compulsory state education was introduced in the 1880s. The chapel choirs were often very good and frequently combined with others to perform such classics as Handel's *Messiah* or

Haydn's *Creation*. Chapel attendance was, if not obligatory, well-regarded, and introduced a note of respectability into what had been the hell-holes of early mining communities. An offshoot was a strong temperance movement as a reaction against the widespread drunkenness, and the dissenting churches were an influence on early trade unionism. With the decline in religious observance over the past 40 years or so, many of the chapels have fallen into disuse, been converted into houses, restaurants, or even mining museums, or have been taken over by other faiths. None the less, for 200 years or more, Methodism and its relatives were a major aspect of social life in mining and industrial communities, and the Salvation Army is still justly famous for its social work.

Mining engineers, who design, develop, and manage mines, came from the same community, and knew and understood its people. Even the flint mines of Grimes Graves in East Anglia had someone in charge who was responsible, probably to a tribal leader, for ensuring the production of flint, negotiating 'wages', and deciding where the next 'grave' was to be. Stretching the point a little, we might reasonably say that such people were Britain's first mining engineers, though we only have the archaeological record of their considerable achievements.

The first mining engineers of which we have any knowledge were the mine captains of Cornwall and Devon. These men were literate enough to keep records and were also very experienced miners with a good 'nose' for where ore might be found. On top of that they had to be very tough, and perhaps not slow with their fists when dealing with recalcitrant or troublesome miners. When times became hard in Cornish tin and copper and Pennine lead, or when better prospects beckoned in other parts of the world, such as in the nineteenth-century gold rushes, many of the captains and thousands of miners emigrated.

This accumulation of hard-won empirical knowledge and practical experience continued to be the norm until the work of Agricola. Georgius Agricola was born in Saxony in about 1495, though in the fashion of the times he Latinized his German name. He studied in Germany and in Italy and became a physician in a town in the richest mining district of Germany. In what turned out to be an unsuccessful search for new medicines in the local minerals he amassed great knowledge of mining and produced the first mining textbook, *De Re Metallica*, 'concerning metals', published posthumously in 1556 but widely translated in the following centuries. One such translation, in 1912, was by Herbert Hoover and his wife. One of Agricola's problems had been to invent Latin words for German mining terms, much as the Vatican has to devise them for innovations such as television and mobile phones. The Hoovers had to reverse that and deduce the English versions of Agricola's invented Latin words—not an easy task. Hoover credited Agricola with being the first person to found the natural sciences on observation, not on conjecture.

In the nineteenth century the education of mining engineers was put on a sound footing, partly due to the evolution of mining engineering from a craft skill to a science-based profession, but also because of legislation requiring that a mine manager have formal qualifications in mining technology and law, as well as specified practical experience. Lower grades of management were also required to have appropriate qualifications and experience. For many years aspiring mine managers worked underground for five shifts a week, or even six if extra cash was needed, and then studied in the evenings at places such as Wigan Technical College. This was a hard route, and many of them were tough and determined men.

To meet the demand for qualified people for the British Empire's burgeoning mining industries, the Royal School of Mines was established in 1851, with profits from the Great Exhibition, as Britain's first university-level institution devoted to mining, geology, and related subjects. Camborne School of Mines in Cornwall soon followed, and schools of mines were founded at universities such as Leeds, Sheffield, Newcastle, Cardiff, and elsewhere in the coalfields. Many of the graduates from these universities were local men who would otherwise have studied part time at technical colleges. People who did not have that local connection were generally treated with great suspicion and even hostility.

With the collapse of Britain's coal industry, the transition from Empire to Commonwealth, and the establishment of highly respected mining schools in South Africa, Australia, Canada, and India, the schools of mines in Britain have all closed, except Camborne and the University of Leeds. The Royal School of Mines no longer exists as such but is now part of the Department of Earth Science and Engineering at Imperial College in London.

The decline in mining education paralleled that of the mining communities, and Britain's mining industries are now a shadow of their former selves. There are only five deep coal mines, whereas once there were 1,600. There are currently no working mines in Cornwall and west Devon, though that may change: as we shall see in the next chapter, two mines may re-open.

There are numerous mining museums and attractions, up and down the country, many getting large numbers of visitors, and offering insights into what mining life was like, often with guided tours underground, and with other events. While many of these museums cater purely for the heritage and tourist market, others are major resources for research and hold archives of mining literature. The national coal museums in England, Scotland, and Wales are notable in this respect. However, museums, successful though they are, can provide only limited employment, often seasonal, and cannot replace the employment of many thousands of men.

In an attempt to offset the effects of mine closures, national and local government, and the communities themselves, have tried hard to develop new sources of employment and income. To take just one case, Elsecar Colliery, near Rotherham, was once owned by the Fitzwilliam family, whom we discussed in Chapter 12. It operated from about 1750, closing in 1984 after the miners' strike. After a long period of disuse, the restored historical buildings now house an antique centre, individual craft workshops, and exhibitions of Elsecar's past. Some parts of the country have gained completely new industries: Sunderland has a large car factory employing some retrained miners, though many of them have had to move from their old communities so as to be nearer to their work. Miners are, however, adaptable and resourceful people, and some have been able to use their redundancy payments to start small businesses, often with help on accountancy and other business matters, though such ventures have their own risks. However, for many former miners, working in a factory might seem little better than drudgery. The camaraderie of a working team, the awareness of danger, and the notion that mining is work for 'real men' cannot be matched, in their eyes, by a factory or a postal round.

As a generalization it is the younger miners who can more easily be retrained or who can take that risk with redundancy money, while older men are all too often left unemployed. Similarly, older miners are more likely to have been affected by industrial diseases of the lungs, and of the hands through the use of heavy, vibrating machinery. The government provided £7.5 billion to compensate them, but there have been scandals, allegedly involving improper handling of that money, and the miners have received less than they had hoped for.

The loss of mining employment hit hardest in the coalfields, the valleys of South Wales being among the worst affected, partly because they are deep valleys and developing lateral communication between them to support new industries is very difficult. Local authorities and, more recently, the Welsh Assembly, with funds from the European Union's regional development budget, have tried hard to get small businesses started and to exploit the area's heritage and natural beauty. Local educational institutions provide training courses. Despite these efforts the fact remains that, while Cardiff is a vibrant, exciting city with a seemingly never-ending building boom, the train ride from there to Aberdare is a sad one. Aberdare has a conference and heritage centre based on an old colliery, but Mountain Ash, which was once a hive of industry with collieries employing thousands of skilled miners, looks depressed and depressing. In short, the old mining communities have effectively vanished and there has been much social upheaval. The chief loss is the community spirit referred to earlier. The gain is that miners no longer toil as they once did.

Yet miners are resilient people. For instance, when British Coal, the eventual successor to the National Coal Board, wanted to close Tower Colliery, the last deep coal mine in South Wales, in 1994 on the grounds that it was uneconomical, the 240 miners pooled their £8,000 redundancy money and bought the pit. They operated it profitably until 2008, when coal reserves finally ran out.

14

Epitaph or Revival? The Possibilities for Britain's Mining Industries

From very ancient times, and even more so since about the sixteenth century, the mining people and their work have shaped Britain's landscape, material prosperity, and way of life, and this book has sought to tell that story. Mining museums and the legacy of mining in the landscape show the passing tourist or the interested visitor how people lived and worked in these once-great industries. And they *were* great. As we have seen, in the latter half of the nineteenth century, Britain's collieries provided 90 per cent of the world's international trade in coal; Britain was a major source of copper in the late eighteenth century; and Welsh slate roofed important buildings throughout the world. Coal mining, though currently successful, is only a shadow of what it once was, and the sites of some former coal mines are now heritage centres, business parks, shopping centres, and sports grounds. Similarly, lead and zinc have not been produced in any serious quantities for many years, and the last tin mine in Cornwall closed in 1998. The Lake District, which once hummed and smoked with mining activity, now delights the visitor with its beauty and peace.

On the face of it, British mining has had its distinguished day—in conversations people have even said that they did not know that the country still had *any* mines and quarries—but that is not the case. We cannot see this book as British mining's epitaph and simply write THE END. To get a truer picture we need to look at the present state of the country's mines and quarries and ask whether British mining has a future, especially as global mining is booming, with serious shortages of the necessary engineers and of skilled and experienced miners. It is difficult to say confidently that British mining has a bright future, but it is not dead, and there are some positive signs.

The current mining and quarrying industries are larger than one might think. Figures from the UK Office for National Statistics for 2007 show that Britain's extractive industries now employ about 30,000 people who produced a total of 293 million tons of minerals. Limestone accounted for 84 million tons, a good deal of which went for concrete, construction generally, and roadstone. Scotland's output of limestone was small, but substantial amounts came from Wales and there was major production in the English Midlands, Yorkshire, and the Mendips. Slate, which includes 'roofing stone', came from Wales, several parts of Scotland, northern England, and Devon, and amounted to 1.4 million tons, undoubtedly influenced by the requirement of planning authorities that new buildings, and repairs to old ones, conform to local tradition, such as the stone roofs of the Cotswolds. Igneous rocks such as granite amount to a further 50.7 million tons, more than half of which comes from Scotland, and is mainly used for roadstone, concrete, and rail ballast. It would be tedious to list too many such cases, but it is surprising how big the UK's extractive industries (excluding oil and gas) still are. China clay is still produced in significant quantities in Cornwall and Devon, although production of clay for coating paper was recently transferred to Brazil, with the loss of several hundred jobs. Coal and salts are worth looking at for the ways in which they have used the most modern technology to maintain profitable working.

Coal was once a huge industry, still employing 700,000 men in 970 collieries when the industry was nationalized in 1947. Nowadays, coal mined in Britain has to face stiff competition from cheaper imports and substitutes such as gas, and there are only a few deep coal mines, though it is possible that more might open, and there are several open-cast operations. Open-cast mining used to be environmentally unpleasant, but great efforts are now devoted to minimizing disturbance and to restoring the landscape when work is completed.

When the coal mines were privatized in 1994 only 15 deep mines remained. The English mines were bought by RJB Mining and are now owned by UK Coal, which describes itself as a mining, property, and energy company, and is the only coal company in Britain operating both deep and surface mines. In all they produce about 8 million tons a year, or half of Britain's total, though 16 million is a far cry from the 200 million tons of 100 years ago. On the other hand, UK Coal employs only about 3,200 people, so productivity is now much higher than it once was. One of the secrets of that is investment in new technology which has enabled a coal face at Daw Mill Colliery near Nuneaton to produce up to 3.2 million tons in one year; another is better industrial relations after the 1984 strike. The company also emphasizes continuing training.

The coal seam, known as the Warwickshire Thick, is about 17 feet high, and the machine is an Eickhoff shearer (plate 27), a major development of the Anderton machine that we mentioned in Chapter 6, but imported from Germany. It runs on top of an armoured conveyor and the cutting drums are one metre wide. The drum that can be seen is cutting the top half of the seam, while a trailing drum cuts the rest. One man, fully equipped with safety gear, controls the machine; this is a highly skilled job. The right-hand side of the photograph shows the roof supports, or chocks, that completely cover the coal face, but fold back as the cutter passes. As the face advances the chocks 'walk' forward under hydraulic power at very high pressure, which also supports the roof. The electrical and mechanical engineers who keep the system running have very demanding jobs, as prolonged inactivity of the shearer risks roof pressure overcoming the hydraulics and jamming the equipment solid. Naturally, the capital cost of this is enormous, as are the problems of taking all those very heavy components down the pit and installing them at the face.

In addition to mining, UK Coal has considerable land and property investments, amounting to some £400 million. The company develops and regenerates brownfield sites, manages business parks on former mine sites, and runs a substantial agricultural portfolio of land and buildings. Only one non-executive director of UK Coal is a mining engineer; all the executive directors come from senior posts in various business areas. The rest of the country's output comes from half a dozen companies which between them operate open-cast mines in England, Scotland, and Wales, and a couple of deep mines in England. New mines may open in Wales—they would be the first new collieries for 30 years—but they are expected to be small drift mines, not the large, deep collieries of the Welsh past.

At present, power stations consume large quantities of coal, much of it imported, but a possible switch to natural gas or the redevelopment of nuclear power could affect the revival of the indigenous coal industry. Some people object to new coal-fired power stations on what they see as environmental grounds, but barring a major programme of nuclear power stations or a breakthrough in renewable sources, coal's place as an energy supplier is likely to be secure for some time to come. However, public policy needs to take into account energy security, so how much coal should come from imports as opposed to local sources is an unresolved issue.

Britain's coal industry is now regulated by the Coal Authority which was established by Parliament in 1994 with statutory duties. One is to license coal-mining operations in Britain, but the authority also has to cope with some of the National Coal Board's responsibilities that were not taken on by the private companies. These relate to subsidence claims, liability for minewater discharged from old collieries, and miscellaneous property matters. In addition the authority provides public information on past and present coal mining.

Britain still has considerable reserves of coal: some 2.3 billion tons suitable for underground mining, and a further 800 million tons mineable from the surface. Most of this is steam coal, but some is coking coal. A relatively novel technology is underground gasification in which coal is turned into gas which can be pumped to the surface. The potential of this is being evaluated, and time will tell, but when the technology is mature the potential reserves from bore holes into old coal mines can only be described as huge.

The coal industry does not stand still, and there is considerable research into new technologies. A significant issue is global warming and the pressure to reduce carbon emissions and to find reliable means of capturing and storing carbon dioxide. The UK government has announced that new coal-fired power stations will have to use carbon-capture technology 'when it is available'. Allied with that are efforts to minimize the production of pollutants such as particulates and sulphur and nitrogen oxide gasses. Finally, there is study of more efficient combustion technologies which would, of themselves, contribute to the two other research aims.

The salt industry prospers—we described Boulby mine in Chapter 8—but Winsford salt mine in Cheshire shows what can be done in a large mine. Winsford produces about 1 million tons per year of rock salt for use on the roads. The rock salt is brown because of the sand mined with the salt. The company is part of the Salt Union and has the unofficial motto, 'De-icing is our business'. The mine uses a room-and-pillar method (the rooms are 9 metres high and 20 wide) that extracts 30 per cent of the salt. The mining machine is a Joy Continuous Miner, imported from Pennsylvania, which with its attached conveyor weighs 130 tons, costs £2 million, and has a working life of 10 years at most. The salt bed is 18 metres, nearly 60 feet, thick, and is mined in two 9-metre slices.

Winsford has five square miles of underground workings and, as well as the pillars between the workings, large ones have to be left to support the River Weaver and the London–Glasgow railway. The tunnels in the mine run for a total of 138 miles; the main roads even having traffic lights at intersections. There is an underground canteen and offices. Worked-out areas of the mine are used to store toxic waste from incinerators, but this required an environmental impact assessment of the geology for the next 50,000 years to show that there was no risk from earthquakes or even volcanoes. So far, more than 40,000 tons have been stowed away, though the eventual total will be 2 million tons. Because the air quality down the mine is as good as it is in the Cheshire fields above it, with a constant temperature of 14°C, other areas are used to store police, company, and medical records, together with parts of the National Archives, with more than 2 million barcoded cardboard boxes stored so far. The locations of the waste- and record-storage areas are secret.

As to the metals, no copper, tin, lead, zinc, iron, tungsten, or manganese is currently produced, though there are probable reserves of copper and other metals on Anglesey, and tin in Cornwall. To explain why these are 'probable' reserves, mining companies have to be very cautious about their ore reserves, the value of which can affect the share price. 'Proved' reserves are those being mined currently, or in the recorded past, and in which the metal content is well understood. 'Probable' reserves are blocks of ore which have been explored by drilling or tunnelling, and the metal content of which can be estimated with reasonable accuracy. Finally, 'possible' ore reserves are estimated from geological factors, or from a neighbouring mine, and are much less certain. A mining company tries to keep a balance between these three categories, as there is no point spending a lot of money on exploration of ore bodies that will not be mined for many years to come; on the other hand, they want enough proved and probable reserves to sustain the forward plan, provided that metal prices and operating costs are viable.

To see the effects of prices, South Crofty tin mine in Cornwall closed in 1998 because of high operating costs and the low price of tin; the reserves were proved but unprofitable. However, in recent years the price of tin has rocketed, so work is being done to widen old tunnels to accommodate modern machinery, in preparation for South Crofty to re-open when circumstances permit. Similarly, tungsten and tin were first discovered at Hemerdon Mine near Plympton in Devon in 1867. The mine operated sporadically until 1944, at which time it closed, but large deposits of tungsten ore were known to remain. Between 1976 and 1980 a major programme of drilling, exploration of the ore body via an adit, and test processing of 6,000 tons of ore indicated about 40 million tons of tungsten and tin ores. The whole ore body may contain 73 million tons. At 1980 metal prices Hemerdon was not viable but, as with South Crofty, it may re-open in an era of higher prices.

As far as our need for metals is concerned, the British Geological Survey (BGS) has estimated that each person, over a lifetime, uses about half a ton of copper, a little less of lead and zinc, very little tin, and about 17 tons of steel, to take just a few examples. Allowing for the population and life expectancy, each of these figures has to be multiplied by about a million to get a rough idea of annual usage, which has to come from home production or imports. Imports can be uncertain and subject to political factors, so there may be some potential for a revival of British metalliferous mining. Cornwall, Devon, and Anglesey have substantial reserves, and the ore deposits of Scotland were evaluated by the BGS 90 years ago. In addition, the BGS conducts geological exploration. Very advanced techniques of geochemistry have suggested the existence of lead and copper ore bodies under Precambrian rocks in Pembrokeshire and under Old Red Sandstone—a thick sequence of Devonian rocks formed between 416 and 360 million years ago—in the Midlands. The BGS also produces a series of Mineral Reconnaissance Reports.

An alternative approach is to study mining areas in detail, and in the 1960s the UK Department of Trade and Industry asked the late John Trounson, a mining engineer with an encyclopaedic knowledge of Cornish mining history and geology, to review mining prospects in a sample of 50 mines which had ceased production for reasons such as inadequate capital, ownership disputes, insufficient pumping capacity, etc. Trounson concluded that many good prospects remained. To take but one case, he writes of Wheal Vor, a mine in west Cornwall, that it has 'so much unexplored ground in an area of exceptionally rich lodes'. In 1971 the Department decided that the reports were no longer required, but this method of detailed examination of historical records by an expert in a given area's mining seems to be a useful complement to the work of the BGS. It should be pursued so that we could at least know what mineral resources Britain still possesses.

However, while it seems that there are undiscovered reserves of copper, tin, lead, and zinc in Cornwall and west Devon, and there may still be lead and zinc in the Pennines and Lake District, whether the quantities are commercially viable is a moot point. A major difficulty is that old mines can be full of water, and pumping that out—referred to as 'de-watering'—is a big task. I was taught to use a ridiculous unit, the acre-foot—which is the amount of water that covers one acre to a depth of one foot (just over 1,233 cubic metres), and in a very wet mine, the inflow can be several acre-feet per day. For comparison, my household water consumption for six months is 35 cubic metres, and 1,200 cubic metres would flood a typical British suburban garden to a depth of two metres. On top of that there is the added problem of bringing old shafts and tunnels up to modern standards of access, safety, and ventilation, and skilled and experienced miners would not now be easy to find.

In addition to the technical problems there is the social issue that areas in which minerals might be found are in National Parks, Areas of Outstanding Natural Beauty, or Areas of Special Scientific Interest, which would arouse much opposition to a mining venture. Indeed, much limestone is quarried in North Yorkshire and the Mendips, and any proposed expansion of quarries is usually bitterly contested by local people. Another illustration is Dartmoor, which, as mentioned earlier, is a gigantic mass of granite, covering 368 square miles, which intruded into the existing rocks about 280 million years ago. The intrusion altered those rocks so that Dartmoor is surrounded by a ring of baked rock, or a metamorphic aureole, in which manganese was once mined, and there may be other minerals there. The trouble is that the aureole coincides with the boundary of Dartmoor National Park, and mining would be unlikely to be popular, despite the jobs and economic benefit it might bring. It is unfair to label this opposition as 'not in my back yard'. The National Parks, Areas of Special Scientific Interest, etc., are there for a purpose, but a balance must be sought between recreation and supply, and mining companies might see little incentive to invest very large sums of money in such uncertain prospects when the pay-off for exploration in, say, Canada might be much greater.

Uranium is something of a special case because, while so-called renewables such as wind farms, some of which are in National Parks and Areas of Outstanding Natural Beauty, might provide 20 per cent of our energy, they still need power stations as back-up, and this is seen as a reason to revive nuclear power. That needs uranium, of which Canada, Australia, and Kazakhstan are, at present, the major producers, with Russia, Namibia, Niger, and Uzbekistan providing lesser quantities. However, Kazakhstan continues to increase production and may become the world's largest producer, exceeding both Canada and Australia. Those two nations will have their own needs and other customers, so opening uranium mines in the Orkney islands (or re-opening Cornish uranium mines) might be advisable to make the UK more self-sufficient in uranium. However, uranium mining in the Orkney Islands was rejected in 1979 by the local planning authority and is likely to be vigorously resisted.

There have even been apocalyptic warnings, not entirely fanciful, of wars being fought over resources. For instance, in May 2009 Russia produced a strategic review, endorsed by the country's president, that identified the potential oil and gas resources of the Arctic as a matter of such concern to Russia that military measures might be necessary. The other countries bordering the Arctic are the USA, Canada, Norway, and Denmark, all of which are NATO members. Britain has no direct interest in the Arctic but perhaps the country ought not to be too dependent on overseas sources for the materials that are essential to modern life. We are fortunate that the BGS has produced numerous reports on our potential resources. However, while geology does not change, the economic and political goalposts move from time to time, especially in the light of predictions about global warming. If those warnings turn out to be true, or even become accepted as true, the demand for resources to build renewable energy sources would increase. Perhaps, therefore, it would now be appropriate to have a rational, and national, debate on where and whether further mining should be permitted, though it is hard to be very optimistic about the future prospects for mining in England, Scotland, and Wales, and the glory days will not return.

We have seen how our present lives and the landscape have been shaped by centuries of toil and hardship by Britain's miners and their families, and the innovation and enterprise of engineers and mine owners. Perhaps in time the remnants of Britain's mining past will become covered in grass (or business parks) and disappear from sight. Even so, we will still have the rich historical record, and the importance of mining to Britain's history and its heritage will not be forgotten.

APPENDIX

The Main British Minerals

This appendix gives you a ready-reference, in alphabetical order, to the main minerals found in Britain and to others that are mentioned in the book. It gives the common name, the main uses of the mineral, or the metal that can be extracted from it, and where it has been mined or quarried in Britain. The list given below is by no means exhaustive, and describes only the principal minerals, as well as some of the minerals that do not produce metals. Many minerals are very beautiful.

This appendix draws, with grateful and affectionate acknowledgment, on one of my university textbooks, *Rutley's Elements of Mineralogy*, 24th edition, edited by H. H. Read, Professor of Geology at Imperial College. It is not clear when Frank Rutley first wrote *Elements of Mineralogy*, but the 19th edition was published in 1915, while the First World War was raging. The 23rd edition was reprinted in 1942 and 1944, in the midst of the Second World War, so it is a tribute to civil servants in those trying times that scarce resources of manpower, paper, and ink were set aside to produce this, and many other, textbooks. Education could not stop just because there was a war on, but the war meant increasing demands for metals and the people to mine them, though some mining engineers served in wartime in the Royal Engineers.

Appendix: The Main British Minerals

MINERAL	USED FOR	COMMON ORES	CHEMICAL COMPOSITION	APPEARANCE	HISTORICALLY MINED IN BRITAIN AT
Alum	Before synthetic dyes were developed alum was used to make brightly coloured dyes	Potash alum, kalinite	Potassium aluminium silicate $KAl(SO_4)_2$	Brightly coloured	Whitby and Isle of Wight
Anhydrite (see also gypsum)	Manufacture of fertilizers, plasters, and sulphuric acid	Anhydrite	Calcium sulphate $CaSO_4$	White with a grey, blue, or reddish tint	Lakeland
Antimony	Used in alloys such as anti-friction materials	Antimonite	Antimony tri-sulphide, Sb_2S_3	Lead-grey	Cornwall

(continued)

233

MINERAL	USED FOR	COMMON ORES	CHEMICAL COMPOSITION	APPEARANCE	HISTORICALLY MINED IN BRITAIN AT
Arsenic	Because it is poisonous, most is used in insecticides and similar products. Also used to harden lead in batteries, and in semiconductors	Mispickel, Arsenopyrite	Iron, sulpharsenide, FeAsS	Tin-white	In tin mines in Devon and Cornwall
Barium	Manufacture of white paint, and production of wallpaper. Also used in medicine for X-rays of the bowel	Barytes	Barium Sulphate, $BaSO_4$	Flat crystals, colourless or white, often tinged with yellow or red. Can occur in masses	Very commonly associated with lead and zinc veins in Derbyshire and Durham

Appendix: The Main British Minerals

MINERAL	USED FOR	COMMON ORES	CHEMICAL COMPOSITION	APPEARANCE	HISTORICALLY MINED IN BRITAIN AT
Bismuth	Used in medicine and for fusible plugs in automatic fire sprays	Bismuthinite	Bi_2S_3	Lead-grey	In veins associated with lead and copper in Cornwall and Cumbria
Calcium (calcium does not occur in pure form, but its numerous minerals are very abundant)	Many uses depending on purity. Poorer quality made into cement, purer varieties used in many industrial processes such as manufacture of soap, bleaching powder	Calcite (see limestone as a separate heading)	Calcium carbonate, $CaCO_3$	Finely developed crystals, and stalactites and stalagmites in caves	Mainly quarried in most limestone areas

(continued)

Appendix: The Main British Minerals

MINERAL	USED FOR	COMMON ORES	CHEMICAL COMPOSITION	APPEARANCE	HISTORICALLY MINED IN BRITAIN AT
	Finest grade used to make opaque glass. Also a source of hydrofluoric acid. Poorer quality spar is used in steel-making	Fluorspar, Blue John, Derbyshire spar	Calcium fluoride, CaF_2	Beautiful crystals: white, green, purple yellow or blue	Weardale in Durham and Castleton in Derbyshire. The blue variety, Blue John, found in Derbyshire, makes spectacular ornaments, but is now rare and very expensive
	A component of cement, a fertilizer, and filler in paper, paint, etc	Gypsum	Hydrated calcium sulphate, $CaSO_4.2H_2O$		In London and Oxford clays

Appendix: The Main British Minerals

MINERAL	USED FOR	COMMON ORES	CHEMICAL COMPOSITION	APPEARANCE	HISTORICALLY MINED IN BRITAIN AT
China Clay	Making fine china and porcelain, and as fillers in paper, rubber, and paint manufacture	Kaolin, china clay	A complex alumino-silicate $Al_4Si_4O_{10}(OH)_8$	White when pure, grey and yellowish	Cornwall
Coal	Domestic heating (now on a small scale), steam and electricity generation, sometimes for black jewellery	There are many types of coal, the most important being bituminous coal and anthracite. Jet is a hard form used for jewellery and found near Whitby	Carbon, about 85–90%, with varying amounts of hydrogen, nitrogen, and oxygen	Generally shiny-black	Very widely in Britain, notably southern Scotland, South Wales, Cumbria, Lancashire, South and West Yorkshire, Nottinghamshire, Leicestershire, the Forest of Dean, and Kent

(continued)

Appendix: The Main British Minerals

MINERAL	USED FOR	COMMON ORES	CHEMICAL COMPOSITION	APPEARANCE	HISTORICALLY MINED IN BRITAIN AT
Copper (many copper minerals occur in Britain, but only the most important ores are mentioned)	Copper has many uses. It is an excellent conductor of heat and electricity, second only to silver for electricity, and is of great importance for any electrical appliance and in heating systems	Cuprite	Copper oxide, Cu_2O	Crystals, sometimes in massive form, different shades of red	Principally in Cornwall, west Devon, Anglesey, and at Great Orme near Rhyl in North Wales. Some of the mines in Cornwall and North Wales are now museums and visitor attractions
	Copper is valuable for alloys such as bronze and special bronzes such as manganese bronze	Chalcopyrite	Sulphide of copper and iron, $CuFeS_2$	Brass-yellow crystals	

MINERAL	USED FOR	COMMON ORES	CHEMICAL COMPOSITION	APPEARANCE	HISTORICALLY MINED IN BRITAIN AT
	Copper chloride, $CuCl$, is used as a disinfectant and in the chemical industry, and copper sulphate, $CuSO_4$, is used to prevent rot in timbers	Chalcocite (a very valuable ore)	Copper sulphide, Cu_2S	Blackish or lead-grey crystals	
	Copper has so many uses that it is fair to say that civilization could not exist without it	Malachite	Hydrated copper carbonate, $CuCO_3.Cu(OH)_2$	(Massive form), bright green with often different shades of colour in concentric bands. A particularly beautiful ore	
		Azurite	Hydrated copper carbonate, $2CuCO_3.Cu(OH)_2$	Deep azure blue, another very beautiful ore	

(continued)

Appendix: The Main British Minerals

MINERAL	USED FOR	COMMON ORES	CHEMICAL COMPOSITION	APPEARANCE	HISTORICALLY MINED IN BRITAIN AT
Gold	Jewellery, high-quality bookbinding, furniture decoration, and some specialized electrical applications	Native gold, or gold alloyed with silver	Gold, Au	Usually as grains or flakes, occasionally as nuggets in alluvial, river gravel, deposits. Grains and flakes are also often found in alluvial deposits	Relatively small amounts have been mined in Cornwall, North Wales, Sutherlandshire, and Leadhills (Scotland). By contrast, there are huge gold mines in Australia, South Africa, the USA, and elsewhere
Graphite	Pencils and art materials, crucibles, paints, electrical applications, stove polish	Plumbago, black lead	Carbon, C, of varying degrees of purity	Black. As nodules in a vein	Borrowdale in Lakeland

Appendix: The Main British Minerals

MINERAL	USED FOR	COMMON ORES	CHEMICAL COMPOSITION	APPEARANCE	HISTORICALLY MINED IN BRITAIN AT
Gypsum (see also anhydrite)	Manufacture of cement, fertilizer, paper, etc.; plaster of Paris and for building plasters and stucco work	Gypsum	Hydrated calcium sulphate, $CaSO_4.2H_2O$	Colourless or white	Lakeland, Leicestershire, Nottinghamshire, Staffordshire, east Sussex
Iron	Iron, and the steel made from it, have an incredible range of uses: everything from zip fasteners to supertankers, via jet engines, railway lines, steel-framed buildings, beautiful iron gates and much, much more. Take a look round your home and workplace and try to count the uses of iron and steel that you see. You may well lose count!	Haematite (also called kidney ore)	Iron oxide, Fe_2O_3	Lumps with a vague resemblance to kidneys, steel-grey to iron black	Haematite occurs in Cumbria and the Forest of Dean. Limonite and siderite were mined in Cleveland, South Wales, and south Staffordshire. Iron pyrites is mainly valued as a source of sulphur

(continued)

MINERAL	USED FOR	COMMON ORES	CHEMICAL COMPOSITION	APPEARANCE	HISTORICALLY MINED IN BRITAIN AT
		Siderite	Iron carbonate, $FeCO_3$	Pale yellow or buff-brownish	
		Magnetite A very valuable ore of iron	Iron oxide, Fe_3O_4	Iron black, strongly magnetic	
		Limonite, brown hematite	Hydrous ferric oxide, approximately $2Fe_2O_3.3H_2O$	Various shades of brown	
		Iron pyrites (fool's gold)	Iron sulphide, FeS_2	Bronze-yellow to pale brass-yellow	

MINERAL	COMMON ORES	CHEMICAL COMPOSITION	APPEARANCE	HISTORICALLY MINED IN BRITAIN AT	USED FOR
Lead	Galena	Lead sulphide, PbS (almost always also contains silver)	Lead-grey cubes	Most localities for lead ores, e.g. Cornwall, Derbyshire, Leadhills (Scotland), Durham, and Cardiganshire	A multitude of uses: batteries, lead sheeting for roofs and for lining processing tanks in the chemical industry, in alloys such as pewter and anti-friction materials, fusible plugs in fire sprinklers, ammunition, pigments, but no longer common in paints
	Cerussite	Lead carbonate, $PbCO_3$	White or greyish, sometimes tinged blue or green by copper.		
	Pyromorphite	Chloro-phosphate of lead, $3Pb_3P_2O_8 \cdot PbCl_2$	Green, yellow, or brown crystals, often very vivid		

(continued)

Appendix: The Main British Minerals

MINERAL	USED FOR	COMMON ORES	CHEMICAL COMPOSITION	APPEARANCE	HISTORICALLY MINED IN BRITAIN AT
Limestone	Strictly speaking limestone is a calcium mineral, but we'll look at it separately because it occurs in colossally thick layers covering great areas. Limestone has large-scale uses for cement-making, in the steel industry, for building, and for road- and railroad-making. A very fine-grained variety of limestone, lithographic stone, can be used in printing. Marble is metamorphic limestone, but is usually of poor quality in Britain	Calcium carbonate	Calcium carbonate, $CaCO_3$ (often with impurities such as sand or flint)	Off-white in layers of enormous thickness and visible in large cliffs	Quarried extensively and on a very large scale in many parts of Britain. Laid down at various stages in Britain's history (see Chapter 2), each of which has its own characteristic fossils

244

Appendix: The Main British Minerals

MINERAL	USED FOR	COMMON ORES	CHEMICAL COMPOSITION	APPEARANCE	HISTORICALLY MINED IN BRITAIN AT
Manganese	Once used in glass-making but principally in the steel industry for making very hard alloys: ferro-manganese and manganese bronze engine parts. Also used in the chemical industry and for making disinfectants	Pyrolusite	Manganese dioxide, MnO_2	Iron-grey	Never a major mineral but mined in west Wales and Devon. Manganese nodules are widespread on the floors of the world's oceans and may one day be very valuable if they can be raised to the surface

(continued)

Appendix: The Main British Minerals

MINERAL	USED FOR	COMMON ORES	CHEMICAL COMPOSITION	APPEARANCE	HISTORICALLY MINED IN BRITAIN AT
		Manganite	Hydrous manganese oxide, MnO(OH)	Iron-black	
Nickel	Nickel has many important uses, such as in coins. It is used in storage batteries. Its main use is in alloys, especially nickel-steels which have many applications, one of which is in jet engines	Kupfernickel, niccolite	Nickel arsenide, NiAs	Pale copper-red.	South Wales
		Nickel pyrites	Nickel sulphide, NiS	Brass-yellow to bronze-yellow	

Appendix: The Main British Minerals

MINERAL	USED FOR	COMMON ORES	CHEMICAL COMPOSITION	APPEARANCE	HISTORICALLY MINED IN BRITAIN AT
Potassium	Potassium has major uses for making fertilizers and in the manufacture of explosives	Sylvite	Potassium chloride, KCl	Colourless or white	Boulby Potash Mine, Cleveland
Salt	Used in cooking and in many industrial processes, such as glass-making, soap-making, etc.	Rock salt, common salt, halite	Sodium chloride, NaCl	Colourless or white when pure	Cheshire
Silica	Has many uses. A *few* are: sand in cement; making computer chips; sandpaper and other abrasives; and pottery.		Silicon dioxide, SiO_2	A variety of colours: pure white rock crystal is used for jewellery; Cairngorm, smoky quartz, is used in Scottish jewellery; flint is silica.	Found throughout Britain in granites; sandstone is silica arising from the decay of granites

(continued)

Appendix: The Main British Minerals

MINERAL	USED FOR	COMMON ORES	CHEMICAL COMPOSITION	APPEARANCE	HISTORICALLY MINED IN BRITAIN AT
Silver	The main use of silver is for jewellery and ornaments. It is now rarely used for coinage. Compounds of silver are used in medicine, photography, and coloured glass, as in Venetian glass	Native silver	Ag		Silver is commonly associated with lead ores, such as galena. There were medieval silver mines at Bere Ferrers, in Devon, and later mines in mid Wales
		Argentite	Silver sulphide, Ag_2S		

MINERAL	USED FOR	COMMON ORES	CHEMICAL COMPOSITION	APPEARANCE	HISTORICALLY MINED IN BRITAIN AT
Silver (*contd*)		Freibergite	Sulphide of copper and antimony, $(Cu,Fe)_{12}Sb_4S_{13}$, but up to 30% of the copper is replaced by silver	Steel-grey to iron-black	It is included here as freibergite occurs in the lead ores of the Pennines and was often mined in preference to common lead ores such as galena for its silver content
Slate	Roof coverings and ornaments and gravestones. The very high quality Welsh slates are exported worldwide	Slate	Slates are formed when soft sandstones, such as mudstone and shale, are affected by metamorphism. There is no simple chemical formula but they are basically silicon minerals	Slate-grey	Quite widespread, notably in north Wales and Lakeland

(continued)

MINERAL	USED FOR	COMMON ORES	CHEMICAL COMPOSITION	APPEARANCE	HISTORICALLY MINED IN BRITAIN AT
Strontium	Red colour in fireworks, and sugar beet processing	Strontianite	Strontium carbonate, $SrCO_3$	Pale green, yellow, grey, and white	Strontian in Aberdeenshire, Weardale, and Yate near Bristol
Tellurium	Used in certain alloys but also in the manufacture of CDs	Tetradymite	$Bi_2(Te,S)_3$, a compound of bismuth(Bi), tellurium (Te), and sulphur.	Pale steel-grey	Very rare in Britain
Tin	The main use of tin is for tin-plate, which is steel with a thin layer of tin, as in tin cans. It also has valuable uses in alloys, such as pewter, solders, bearings, bronze, bell metal, etc	Cassiterite or tinstone	Tin oxide, SnO_2	Usually black or brown	Mainly Cornwall and west Devon.

MINERAL	USED FOR	COMMON ORES	CHEMICAL COMPOSITION	APPEARANCE	HISTORICALLY MINED IN BRITAIN AT
Tin (contd)		Stannite, tin pyrites, also called bell metal ore	Copper, tin, iron sulphide, Cu_2SnFeS_4	Steel-grey when pure, sometimes bronze or bell-metal colour	
Tourmaline	Good-quality tourmaline is used as a gemstone		An exceedingly complex mineral containing boron, which has very many uses but is more easily obtained from other boron minerals, aluminium, and silica, together with sodium, calcium, iron and magnesium	Black or bluish-grey	Veins of tourmaline occur in Cornish china-clay pits

(continued)

Appendix: The Main British Minerals

MINERAL	USED FOR	COMMON ORES	CHEMICAL COMPOSITION	APPEARANCE	HISTORICALLY MINED IN BRITAIN AT
Tungsten	Mainly used to make steel for tools. Tungsten carbide is used in drills that can bore holes in brick or stone—you probably have one in your tool kit at home. Also used for lamp filaments	Wolframite	Tungstate of iron and manganese, (Fe, Mn)WO$_4$; Mn is manganese and W is tungsten from its German name, wolfram.	Chocolate-brown to reddish-brown.	Cornwall (St Austell), Cumbria Devon Caldbeck Fell (Cumbria), Cornwall
		Tungstite	Tungsten oxide, WO$_3$	Earthy, bright yellow	
		Scheelite	Calcium tungstate, CaWO$_4$	Yellowish-white or brownish	

Appendix: The Main British Minerals

MINERAL	USED FOR	COMMON ORES	CHEMICAL COMPOSITION	APPEARANCE	HISTORICALLY MINED IN BRITAIN AT
Uranium	Nuclear and medical uses	Pitchblende, uraninite	Very variable but typically $2UO_3 \cdot UO_2$	Velvet black, greyish or brownish	Cornwall, and occurs on Orkney
Zinc	Zinc has many uses, notably for coating, or galvanizing, iron. It is also used in alloys such as brass. Zinc oxide and zinc sulphate are used in pigments. Other compounds of zinc are used for treating wood, dyeing, glue-making, medicines, and in other processes	Sphalerite, zinc blende, Black Jack	Zinc sulphide, ZnS, the common ore of zinc	Usually black or brown	In most lead and zinc mining areas, such as Cornwall, Cardiganshire, Derbyshire, Mendip Hills, Alston Moor in Cumbria, and Leadhills in Scotland
		Smithsonite, calamine	Zinc carbonate, $ZnCO_3$	White, greyish, or greenish	

BIBLIOGRAPHY

General

Cameron, A. (ed.), *Lakeland's Mining History*, Cumbria Amenity Trust and Mining Research Society, 2000.

Livingstone, A., *Minerals of Scotland—Past & Present*, National Museum of Scotland Publishing Ltd, 2002.

Thomas, T. M., *The Mineral Wealth of Wales and its Exploitation*, Oliver and Boyd, 1961.

Samuel, R. (ed.), *Miners, Quarrymen and Saltworkers*, Routledge Kegan Paul, 1977.

Chapter 1

Butler, C., *Prehistoric Flintwork*, Tempus Publishing Ltd, 2005.

Chippindale, C., *Stonehenge Complete*, Thames and Hudson, 1985.

Cunliffe, B., *Europe Between the Oceans: 9000BC–AD 1000*, Yale University Press, 2008. (A truly marvellous book, beautifully illustrated.)

Mithen, S., *After the Ice*, Weidenfeld and Nicholson, 2000. (A remarkable account of early human societies.)

Pollard, J. and A. Reynolds, *Avebury: The Biography of a Landscape*, Tempus Publishing, 2002. (Avebury is, in some ways, more interesting than Stonehenge, and this is an enthralling account of how Neolithic people shaped that landscape.)

Chapter 2

Gould, S. J., *Wonderful Life*, Hutchinson, 1990.

Gradstein, F. M., J. G. Ogg, A. G. Smith, *A Geologic Time Scale*, 2004, Cambridge University Press, 2004.

Price M. and K. Walsh, *Rocks and Minerals*, Dorling Kindersley, 2005. (Beautifully illustrated.)

Chapter 3

Barton, D. B., *A History of Tin Mining and Smelting in Cornwall*, D. Bradford Barton Ltd, 1967.

Bick, D., *The Old Copper Mines of Snowdonia*, Round House, 1985.

Bullen, L. J., *Mining in Cornwall*, vols 3 to 7, Tempus Publishing, 1999.

Harris, J. R., *The Copper King: Biography of Thomas Williams of Llanidan*, Liverpool University Press, 1964.

Harris, T. R., *Dolcoath: Queen of Cornish Mines*, Trevithick Society, 1974.

Keay, J., *The Great Arc: The Dramatic Tale of How India was Mapped and Everest was Named*, HarperCollins, 2001.

Lewis, C. A., *Prehistoric Mining at the Great Orme*, MPhil thesis, University of Wales, Bangor, 1996; which can be read on www.greatorme-mines.info. (The most detailed source for Great Orme.)

Penhallurick, R. D., *Tin in Antiquity*, Institute of Metals, 1986.

Rowlands, J., *Copper Mountain*, Anglesey Antiquarian Society, 1966.

Trounson, J. H. and L. J. Bullen, *Mining in Cornwall*, vols 1 and 2, Tempus Publishing, 1993.

Bibliography

Chapter 4

Barton, D. B., *The Cornish Beam Engine*, Cornwall Books, 1969, repr. 1989.

Weightman, G., *The Industrial Revolutionaries*, Atlantic Books, 2007. (Not much about mining, but a very useful complement covering all the other areas of innovation.)

Chapter 5

Burt, R., *The British Lead Mining Industry*, Truran, 1984.

Forbes, I., *Lead and Life at Killhope*, Killhope, the North of England Lead Mining Museum (not dated).

Gill, M., *Swaledale, its Mines and Smelt Mills*, Landmark Publishing, 2004. (A very detailed survey.)

Morrison, J., *Lead Mining in the Yorkshire Dales*, Dalesman Publishing Co., 1998.

Raistrick, A. and B. Jennings, *History of Lead Mining in the Pennines*, Longmans, 1965.

Raistrick, A. and A. Roberts, *Life and Work of the Northern Lead Miner*, Beamish North of England Open Air Museum and Northern Mine Research Society, 1984. (Mainly a superb collection of photographs with explanatory text.)

Willies, L., *Lead and Leadmining*, Shire Publications Ltd, 1982, repr. 1999.

Chapter 6

The literature on coal is enormous, including learned tomes, detailed technical reports on individual mines (one of which was written by the author), archaeological reports by mining research, and biographies of leading personalities.

Bibliography

Adeney, M. and J. Lloyd, *The Miners Strike, 1984/5: Loss Without Limit*, Routledge and Kegan Paul, 1988.

Ashton, T. S. and J. Sykes, *The Coal Industry of the Eighteenth Century*, Manchester University Press, 1964.

Bruce, J. Collingwood, *A Handbook to the Roman Wall*, 12th edn, 1966 (page 117).

Down, C. G., *The History of the Somerset Coalfield*, Radstock Museum, 2005.

Galloway, R. L., *A History of Coal Mining in Great Britain*, David and Charles Reprints, 1969, originally published in 1882. (Despite its apparent antiquity it is an excellent source, as Galloway was a qualified mine manager, a competent historian, lived through the times he describes, and knew many of the people involved. A preface brings the tale up to the 1960s.)

Gregg, P., *A Social and Economic History of Britain, 1769–1960*, George G. Harrap and Co., 3rd edn, 1962. (Now slightly dated, but still an excellent and beautifully written coverage of the whole subject. Especially good on coal mining.)

Kynaston, D., *Austerity Britain 1945–51*, Bloomsbury Publishing, 1951. (An excellent account of what life was like, for those who did not live through those years.)

Marr, A., *A History of Modern Britain*, Pan Books, 2008.

The National Coal Board commissioned eminent economic historians to write the *History of the British Coal Industry*, published by Oxford University Press: volume 1: Hatcher, J., *Before 1700* (1992); volume 2: Flinn, M. W., *The Industrial Revolution* (1984); volume 3: Church, R., *Victorian Pre-eminence* (1986); volume 4: Supple, B., *The Political Economy of Decline* (1987); volume 5, Ashworth, W., *The Nationalised Industry* (1986). These vary in readability and are not for the faint-hearted. Church's book, for instance, runs to 789 pages plus 20 pages of bibliography.

Bibliography

Prominent personalities are covered in *Who's Who* for those still alive, and *Oxford Dictionary of National Biography* for those who are not.

Chapter 7

Armit, A., *Towers in the North: The Brochs of Scotland*, Tempus Books, 2002.

Bezzant, N., *Out of the Rock*, Heinemann, 1980. (On Portland and Bath stone.)

Donelly, T., *The Aberdeen Granite Industry*, Centre for Scottish Studies, University of Aberdeen, 1994. (A scholarly account.)

Elms, F. H., *'Tanky' Elms: Bath Stone Quarryman*, C. J. Hall, 1984.

Jones, R. M., *The North Wales Quarrymen*, University of Wales Press, 1982. (A very well written and informative adaptation of a PhD thesis. Recommended for social history.)

McLaren, J., *Sixty Years in an Aberdeen Granite Yard*, Centre for Scottish Studies, University of Aberdeen, 1987. (A very entertaining personal reminiscence, but also very good on the stonemason's tools and techniques.)

Perkins, J. W., et al., *Bath Stone: A Quarry History*, Kingsmead, 1979, repr. 1990.

Pevsner, N., *The Buildings of England*, Penguin Books, various dates. (A separate volume for each English county, describing the stone used in local architecture.)

Richards, A. J., *Slate Quarrying in Wales*, Gwasg Carreg Gwalch, 1995.

Williams, J. L. W., 'W. J. Parry: Quarryman's Champion?', *Llafur*, 8:1 (2000), 97–110. Society for the Study of Welsh Labour History.

Chapter 8

An excellent set of 26 leaflets on the Cheshire salt industry is available from the Salt Museum at Northwich, www.saltmuseum.org.uk.

Bibliography

Miller, I. (ed.), *Steeped in History: The Alum Industry of North-East Yorkshire*, North Yorks Moors National Park Authority, 2002.

Tyler, I., *Gypsum in Cumbria*, Blue Rock Publications, 2000.

Chapter 9

Cleere, H. and D. Crossley, *The Iron Industry of the Weald*, Merton Priory Press, 1995. (A meticulous account, based on detailed research by county archaeologists and the Wealden Iron Research Group, but mainly dealing with iron-working rather than ore mining.)

Owen, J. S., *Cleveland Ironstone Mining*, Tom Leonard Mining Museum, Skinningrove, Cleveland, 1995.

Squires, S., 'The Underground Mines of Lincolnshire', in *All Things Lincolnshire*, eds J. Howard and D. Start, Society for Lincolnshire History and Archaeology, 2007.

Chapter 10

Annals, A. E. and B. C. Burnham, *The Dolaucothi Gold Mines*, University College, Cardiff, 1996.

Hall, G. W., *The Gold Mines of Merioneth*, Griffin Publications, 1977.

Hindley, G., *A Brief History of the Anglo-Saxons*, Constable and Robinson, 2006.

Chapter 11

Miller, J., *Aberfan: A Disaster and its Aftermath*, Constable, 1974.

Chapter 12

Bailey, C., *Black Diamonds*, Penguin Press, 2007. (History of the Fitzwilliam family, but also mine disasters and mining people.)

Bibliography

Harris, J. R., *The Copper King: A Biography of Thomas Williams of Llanidan*, Liverpool University Press, 1964.

Jones, I. W., *Gold, Frankenstein and Manure*, Llechwedd Publications, 1997. (The fascinating correspondence of John Whitehead Greaves.)

Jones, I. W., *The Eagles of Llechwedd*, Llechwedd Publications, 2004.

Mee, G., *Aristocratic Enterprise*, Blackie, 1975.

Chapter 13

Benson, J., *British Coalminers in the Nineteenth Century: A Social History*, Gill and Macmillan, 1980/Longman, 1989.

Beynon, H. and T. Austrin, *Masters and Servants: Class and Patronage in the Making of a Labour Organisation*, Rivers Oram Press, 1994.

Coombes, B. L., *These Poor Hands: A Miner' Story*, Victor Gollancz, 1939, reissued 1974. (See also the *Dictionary of National Biography*.)

Fraser, H. W., *Trade Unions and Society: The Struggle for Acceptance, 1850–80*, Allen and Unwin, 1975.

Harris, J., *Private Lives, Public Spirit: Britain 1870–1914*, part of the Penguin Social History of Britain, Penguin Books, 1994.

Harrison, R., *Independent Collier: The Coal Miner as Archetypal Proletarian Reconsidered*, Palgrave Macmillan, 1978.

Moore, R. J., *Pit-Men, Preachers and Politics: The Effects of Methodism in a Durham Mining Community*, Cambridge University Press, 1979.

Orwell, G., *The Road to Wigan Pier*, Penguin Classics, 2001, first published by Victor Gollancz, 1937.

Perchard, A., *The Mine Management Professions in the Twentieth-century Scottish Coal Mining Industry*, Mellen Press, 2007.

Trevelyan, G. M., *English Social History*, Longmans, Green and Co., 1944. (Good summary on the coal industry, but a masterpiece covering the six centuries to Queen Victoria's reign.)

Bibliography

The Miners in Fiction

There are numerous novels and films depicting the life and times of the mining people. A few classics are given below.

Billy Elliot: Film and stage musical about a miner's son who wants to be a ballet dancer.

Brassed Off: Film about a colliery brass band in the 1984 strike and the effects of the strike in a fictional (but fact-based) pit village.

Drabble, Margaret, *The Peppered Moth*, Penguin Books, 2001.

Cronin, A. J., *The Stars Look Down*, Victor Gollancz, London 1935. Made into a film, 1939.

Hines, B., *Kes: A Kestrel for a Knave*, Longman, 1982.

Lawrence, D. H., *Sons & Lovers*, eds Carl and Helen Baron, Cambridge University Press, 1992.

Chapter 14

Trounson, J. H., *Cornwall's Future Mines: Areas in Cornwall of Mineral Potential*, University of Exeter Press, 1993.

Vaughan, R., *The Arctic: A History*, rev. edn, Sutton Publishing, 2007. (A gem of a book.)

Wilson, G. V., *The Lead, Zinc, Copper and Nickel Ores of Scotland. Special Reports on the Mineral Resources of Great Britain*, volume XVII. HMSO, 1921.

Websites for the British Geological Survey and the UK Office for National Statistics have a wealth of data, usually about one year in arrears.

INDEX

Index

Index

Index

Index